Sharing for Survival

In memory of

Richard Douthwaite 1942-2011

SHARING for SURVIVAL

Restoring the Climate, the Commons and Society

Edited by Brian Davey with essays by Richard Douthwaite
(assisted by David Knight), Justin Kenrick, Laurence Matthews,
John Jopling, Nick Bardsley, Caroline Whyte,
James Bruges and Brian Davey

Designing systems for a changing world

feasta

First published in Dublin (March 2012) by Feasta,
The Foundation for the Economics of Sustainability
Registered office: The Greenhouse, 17 St Andrews Street, Dublin 2
Administrative office: Main Street, Cloughjordan, Co. Tipperary.
info@feasta.org www.feasta.org
Feasta is registered as an educational charity No. 13052

Distributed in Britain by Green Books Ltd., Foxhole, Darlington, Totnes, Devon, TQ9 6EB
Tel: (01803) 863 843 Email: john@greenbooks.co.uk

A CIP record for this title is available from the British Library.
ISBN 978-0-9540510-1-3
Printed in Ireland on FSC-certified paper for JDK Design
Text designed and typeset by Jeff Kay, info@jdkdesign.net
Cover designed by e-Digital Design

Contents

Acknowledgements

Towards the end of writing this book, in November 2011, our fellow author and colleague, Richard Douthwaite, died of cancer. The title for the book, which was decided quite late in the process, was suggested by Richard and he worked on his chapter despite his illness and declining energy, very nearly to the end of his life, at which time he had finished a late draft. Towards the end however he was too weak to follow up with checking some of the up-to-date scientific information and references and he asked David Knight to help. David stepped into the breach and then, at Richard's request, checked the consistency of the rest of the chapter with the material he discovered. We owe a tremendous debt to Richard that he continued working despite his illness – and then to David that he finished the chapter. In turn, David wants to thank Claire Jones for perceptive and thoughtful comments about this chapter. David responded as much as he has been able although, obviously, it is not possible to know exactly how Richard would have responded to some of Claire's comments.

Our thoughts are with Mary, Richard's wife, and also thanks for her suggestions about how we follow up publication.

Many other people helped with this book in one way or another. It has been a collective effort much more than because a number of us wrote separate chapters. Particular thanks are due to Andy Ross, for help in editing.

Dr Fauzia Shariff introduced us to the concept of social transfers and was thus an important influence in Caroline Whyte's chapter while Anandi Sharan, who earlier researched and authored a study of cap and share in India has contributed a statement on the position of the Green Party of India that supplements the chapter by James Bruges. Our thanks are due to both of them.

We also need to express our appreciation to Mike and Glenys Thomas of the Cap and Share Campaign who hosted one of our discussions – it was Mike whose original idea it was to write this book.

Milena Büchs and Nick Woodman are also due acknowledgement for advice and support as are David Healey, Morag Friel, Bruce Darrell and other members of the Feasta Climate Group who arranged and hosted one of our discussions in Cloughjordan in Ireland.

Nor should we forget the late Will Howard who was so instrumental in getting the Cap and Share campaign off the ground – despite knowing that he, too, had a terminal illness.

Many others gave constructive criticism but our failures are, of course, our own…

Contributors

Brian Davey trained as an economist but, aside from a brief spell working in eastern Germany showing how to do community development work, has spent most of his life working in the community and voluntary sector in Nottingham particularly in health promotion, mental health and environmental fields. He helped form Ecoworks a community garden and environmental project for people with mental health problems. He is a member of Feasta Climate Working Group and co-ordinates the Cap and Share Campaign.

Justin Kenrick is a social anthropologist and activist who lives in Portobello, Edinburgh, where he helped co-found PEDAL – Portobello Transition Town in 2005. He lectured in social anthropology at the University of Glasgow from 2001 to 2009, when he left to work for the Forest Peoples Programme supporting Central African forest peoples to protect their rights to their forests. He has been heavily involved in campaigning against nuclear weapons, in campaigns against the logging of Canada's old growth forests, and currently works with a loose network of resilience-building communities across Scotland.

Laurence Matthews is a mathematician who has worked as a university lecturer and in the transport industry, where he carried out consultancy work on five continents. For several years he has lectured and campaigned on the psychology of climate change, and has given evidence to the Environment Select Committee of the UK House of Commons. He is Chair of Cap & Share UK, an NGO promoting Cap & Share, which is described in Chapter 3.

Nick Bardsley lectures in climate change economics at the School of Agriculture Policy and Development at the University of Reading. He is interested in ecological and behavioural economics, particularly deepening understanding of energy "rebound effects" and evidence that contradicts received theories of economic behaviour.

John Jopling. For 30 years John practiced as a barrister in London advising clients about the law of trusts. Increasing awareness of the deep-seated flaws in mainstream economic and political systems led to using his professional expertise to help establish a number of new institutions, including FIELD the Foundation for International Environmental Law and Development and Feasta the Foundation for the Economics of Sustainability. Publications include two Feasta Reviews, edited jointly with Richard Douthwaite, and the Schumacher Briefing "Gaian Democracies", written jointly with Roy Madron.

James Bruges, with David Friese-Greene, is working with Social Change and Development, an NGO in Tamil Nadu, on the Soil Fertility Project. This combines biochar with sludge from anaerobic digestion as an organic alternative to synthetic fertilisers for the 5,000 farmers connected with SCAD. He is author of The Big Earth Book and The Biochar Debate.

Caroline Whyte grew up in Belfast and Dublin. Her involvement with Feasta began while she was living in the western United States. She collaborated with Richard Douthwaite on an online update of his book Short Circuit: Strengthening Local Economies in an Unstable World in 2002-3 and went on to study ecological economics at Mälardalen University in Sweden in 2005-6, writing a masters thesis on the relationship between central banking and sustainability. She lives in central France, from where she edits the Feasta blog.

Richard Douthwaite was an economist, journalist and author specialising in energy, climate and sustainability issues. He was a co-founder of Feasta, served on its executive and co-ordinated the Feast Climate Working Party. His books included The Growth Illusion (1992), Short Circuit (1996) and the Ecology of Money (1999). He also edited and contributed to "When the Wells Run Dry" (2003) and Fleeing Vesuvius" (2010). Until his death in November 2011, he lived in Westport Co Mayo.

David Knight has been a research scientist and teacher in Life and Medical Sciences since 1970 with a particular interest in carbon-neutral biomaterials. His parallel interest in Ecology, Energy Policy and Economics started in 1973 when he and Prof. Paul Smoker at Lancaster University started "Half Life" to oppose civil and military nuclear power in NW England. He is currently an advisor on the science of energy and climate change to both FEASTA and Winchester Action on Climate Change. At Richard Douthwaites request David assisted with natural science references and some editorial work when Richard was too ill to complete his chapter.

Preface: Brian Davey

As the different authors wrote this book it became clear that each of us had very different perspectives on how climate mitigation could be achievable. We all started with a common sympathy for a policy called "cap and share" but even that unravelled to some degree.

To explain: the policy of Cap and Share would mean that a reducing ceiling, or cap, would be imposed on fossil fuel suppliers through a "permit to sell carbon fuels" scheme. The fossil energy suppliers would have to purchase a reducing number of permits. The money raised by the energy suppliers purchase of permits would be available to share with citizens in various ways.

We all started out liking that idea but then discovered that we had different views of how to organise the 'share'. Part of the book is about this difference – and it touches on something rarely discussed in climate policy – how to ensure that climate policy is fair to individuals, households and communities. Issues like this are crucial if climate policy is to gain public acceptance yet they rarely get discussed.

There were other differences that emerged too – but, as the writing progressed it is these differences that are some of the most interesting aspects of the book.

In summary readers will find the following divergences:

What's the target audience for climate mitigation?

Although the book set out to be for a general lay readership it is clear that we nevertheless have different target audiences. For some of us, effectively, the chapters have been written with an assumption that our ideas might reach and influence government and "inter-governmental" policy makers. For example one of the arguments put forward by the late Richard Douthwaite was for governments in oil importing countries to set up an oil importers cartel. This is a macro-economic policy for the period after "peak oil" aimed at managing the energy market in a way that would stabilise the international economy. Richard argued that this would be in the interests of fossil fuel exporters and importers because as depletion proceeds, a spike in the price of oil and gas as depletion proceeded would crash the world economy and the price of fuels would crash down again. His view was that this would be against the interests of exporters too. There is no need to go into the argument further here – the main point here is to point out that this is an argument about what governments might do. By way of contrast, many of the other chapters have been written with scepticism that governments are capable of delivering any kind of climate action, at least at this stage – even where a case can be made out that there is good reason for common action.

What forms for global climate governance and how to arrive at them?

Another viewpoint in these pages is that it is necessary, and hopefully possible, to strive towards a directly established international climate policy regime with new institutions. This is in contrast to the current approach where each government, albeit sometimes working as part of a negotiating block like the EU, offers what it is prepared to contribute in the UNFCCC negotiations. In theory the aggregate global climate mitigation effort will then emerge out of the sum of all the offers added together. The contrasting approach put here is to start out taking a whole world view, set up global world institutions appropriate to this view, and then to work towards governments 'buying into' the global arrangements.

The main actors in this kind of process, and hence the main audience to whom we would be addressing ourselves, are those with the most credibility in global civil society, independent of government and independent of corporations. It is they that must agree, organise and evolve the organisation and policies first and foremost. That said, one must acknowledge that a civil society process can only ever get so far if it does not at some point engage with and enlist the support of governments who would then be able to bestow the authority to impose particular policies.

The contrasting approach is connected to the view that the earth's atmosphere is a global commons and should be managed as such, in the interests of everyone, independently of states and corporations. Out of that would hopefully come something like a global Cap and Share scheme and other global policies to improve carbon sinks, to deal with methane, N_2O, halocarbons, black soot and so on.

But as we wrote divergences emerged here too. As one author, Justin Kenrick put it about the approach just mentioned, this perspective "wants a way to say the global commons is possible and that governance structures flow from that, whereas my sense is that there are only ever local commons which can flow together to overthrow state structures and from that regain the global spread of relocalising commons regimes."

So, Justin Kenrick's perspective is off in a very different direction. His focus is on people's movements and the relationship between communities and their resources in their local environments, developing a climate policy out of that. When we started to write this book there was no such thing as the Occupy Movement that has spread all round the world. Now that it exists different styles of politics and different vehicles for change are becoming visible. A perspective of change from below, perhaps in the context of a generalised banking and financial system collapse, becomes plausible – and it adds to the plurality of perspectives in this book.

How do we organise the share?

Then, as mentioned above, it is clear that there are different 'angles' on the 'share' in Cap and Share. When one of our authors, James Bruges, explained Cap and Share to a group in India they did not like the idea of a per capital share of carbon revenues going to individuals. This has prompted James to write the piece that he has. In complete contrast, another author, Caroline Whyte, points to the experience of giving cash to the poor in 'developing' countries and finds it to be positive.

"Wicked problems"

Even a cursory look at these different points of view makes it obvious why divergences emerged. There's a jargon phrase to describe the climate situation – it is what is called a "wicked problem". Assessments of the degree of seriousness, causes, contexts and solutions for the climate problem have a high degree of uncertainty and contestation. There are multiple stakeholders and interest groups affected in quite different ways and to different degrees. As a global problem there are, and have been, multiple solutions put forward for multiple parts of the problem by multiple actors. The climate crisis is connected in many and complicated ways to many other kinds of problems – such that things that might be helpful as climate mitigation can make other problems worse.

In these circumstances it would surprising if we did not face fragmentated viewpoints. That's what a "wicked problem" is like. It defies easy conceptualisation.

With a caveat spelled out in more detail below, this diagram, called the Stacey Matrix, is quite helpful to clarifying the sort of situation that we are in.

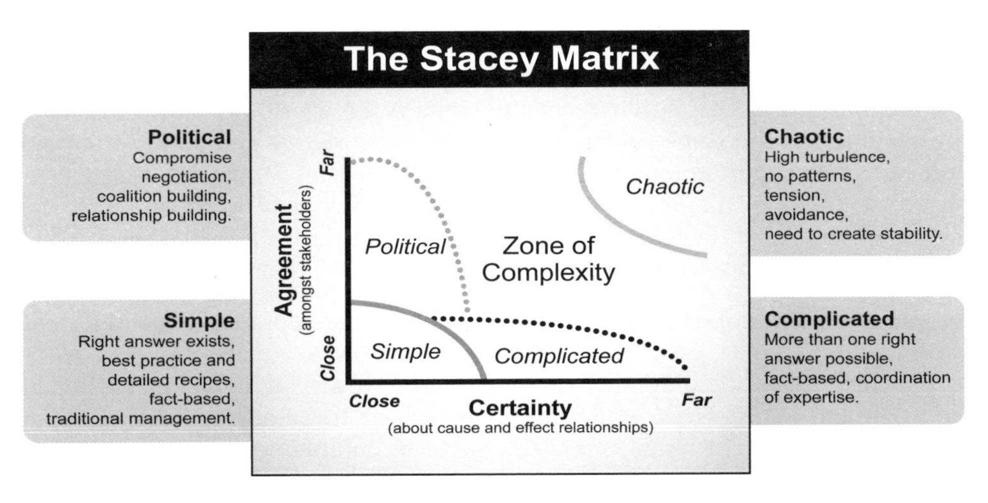

Adapted from: Ralph Stacey Strategic Management Organizational Dynamics: The Challenge of Complexity. 3rd Edition. Harlow: Prentice Hall. 2002.

The climate crisis a problem were there is a great detail of uncertainty – not, I hasten to add about the basic science – but certainly about the contexts in which climate policy is evolving. There is also a good deal of manufactured uncertainty created by vested interests, keen on promoting doubt and denial in order to achieve inaction – which would suit their short term financial interests. At the same time, in regard to the vertical axis there is very little agreement – so the scope for compromise, negotiation, relationship building seems to be low.

We are therefore operating in the "zone of complexity" and neither simple prescriptions, "right answers" about what is to be done about climate change – nor political formula for deals, nor the co-ordination of a huge amount of expertise is enough to come up with "the right answer."

Yet some answer must be found... unless one says, as has author, Clive Hamilton, that the battle against runaway climate change has been lost... but even then Clive Hamilton does not conclude that there is nothing more to be done. (We have had differences on Hamilton too – to Richard Douthwaite Hamilton's view was an anathema. Richard felt strongly that we should be optimistic. Speaking only for myself I'm not that sure, I think our situation is pretty desperate but that people are often at their finest when their backs are to the wall. This however depends on acknowledging how bad things are first).

A patchwork quilt of solutions

Once upon a time, I was a Trotskyite who thought one had to have a unified line which everyone followed in order to achieve fundamental changes in society. I tend to believe the opposite now. I see more use in a patchwork quilt of approaches and solutions, some of which may not match. In a natural crisis, that's how biodiversity works. In the diversity of species with a wide range of characteristics there are some that are able to thrive in a general eco-system collapse. They become pioneer species, starting the evolution of a new system.

In confusing times one approach that makes sense to me is to put lots of seeds out trying lots of different things. The more approaches that are tried the more likely some are to work. As we are not all approaching the climate problem, and the limits to growth situation, from the same place we do not have the same connections, or the same resources, the same knowledge and ability to apply ourselves to a common unified path. It is inevitable that we try different paths. There's a metaphor that is commonly used in discussions of spirituality that the view from the top of the mountain is the same. However because we are starting out from different points, the path up the mountain will be different.

So we should not be afraid of difference. We can compare our different experiments. We can go our different ways and try to create frameworks or maps which relate the different approaches that we are taking to each other. In fact, different approaches, while they may appear to be alternatives between which choices must be made, often turn out to be complementary – as when it is necessary for someone to take an extreme view which hardens the negotiating position for someone who is more prepared to compromise because they see a danger in extreme conflict.

Chapter 1

What can be done if mainstream politics loses interest in climate change

Brian Davey

After Copenhagen it was by no means obvious that simply calling upon governments to act would achieve very much. Yet the situation is urgent - so what do we do? The aim of this chapter is to look at options from getting from where we are now to adequate climate mitigation. It starts by looking at all the obstacles to getting things done - but this is not so that we get discouraged and give up. It is so that we are realistic and can find our way around the obstacles.

A recent book by the Financial Times columnist and academic, John Kay, points out that the most successful ways of achieving policy, business or other goals in human affairs is not to approach our goals directly but indirectly. It is the oblique approach that often achieves most[1].

There is a very good case for approaching climate mitigation obliquely particularly as the task is huge and complex, because much of what needs to happen is unclear – and because the resistances to getting action put in place by vested interests are very powerful. At the same time there are powerful pressures to get something done about a growing crisis in the energy system and millions of people are having to adjust their lives to this energy and economic crisis. So how can an indirect response to the climate crisis be put in place as part of a general programme for the wider crisis? How can we enlist the active involvement of millions of people and win them over for adequate climate policies – for example those who have become involved in the Occupy movement that has sprung up the world over?

If it is not as easy as it is supposed to be to make the democratic process work for us perhaps this is because we have pinned our thinking too much to the head on direct route. People are struggling to cope with lots of problems – how about ideas about how to help them and deal with the climate crisis too?

Most of us know the head on direct route very well. It is the route of political common sense. We are supposed to put credible policy ideas into letters and articles for newspapers and in the letters that we write to our MPs. Having convinced our MPs what is supposed to happen is that our ideas are passed on to ministers and examined by officials. If enough members of the public want something the policy will eventually be enacted. What we are supposed to do is to lobby the politicians and officials with credible ideas. That is the theory and most of us know in our hearts that it doesn't work – even if we do not acknowledge it yet in our heads and in what we do and say.

As if!

As is very clear the chances of getting adequate climate change mitigation in the current growth economy are very slim. The UK government's former advisers, the Sustainable Development Commission, have published studies that say so. For example, "Prosperity without Growth" written by Professor Tim Jackson, showed how a growing economy could not possibly achieve the carbon emissions reductions required even to reach an inadequate 450ppm CO_2 target by 2050. To achieve an average year on year reduction of emissions of 4.9% with 0.7% population growth and 1.4% income growth would require technological change to reduce emissions per unit of economic output at 7% per annum. That is ten times the current rate.[2]

Nevertheless the policy makers and business are locked into a commitment to growth. Growth is a central idea in what John Jopling and Roy Madron term "the elite consensus" in their book on *Gaian Democracy*. [3] Those people who argue for non growth economics are ignored by policy makers, business and most journalists. The Sustainable Development Commission and Tim Jackson told the government that growth and sustainability were not compatible – and this probably helped to seal the fate of the SDC – it was abolished by the coalition government as one of the victims of the cuts.

If you follow the route of political common sense and lobby for ideas outside the elite consensus – ie the growth consensus – you get ignored. Although everyone says that they like thinking that is "outside the box" – they do not mean thinking outside the growing economy box.

Now there are systemic reasons for this addiction to growth. There are reasons as to why it is considered more important than dealing with climate change. For one thing growth has come to be seen as "the" answer for all political problems. Writer Clive Hamilton describes this as fetishistic:

"Growth alone will save the poor. If inequality causes concern, a rising tide lifts all boats. Growth will solve unemployment. If we want better schools and hospitals then economic growth will provide. And if the environment is in decline then higher growth will generate the means to fix it. Whatever the social problem, the answer is always more growth" [4]

Over and above the fetishist mind set of the policy establishment there are deeper, structural, reasons for their collective fixation. These reasons arise out of the nature of the money and financial system. The argument here is not new – green economists have called attention to this problem for decades and it is explored in the other chapters of this book at length.

Debt based money and growth

Since we all depend on the smooth functioning of the money and financial system, and since we all use it in our everyday life, the money system should be regarded as a commons resource. It should be managed in the interests of everyone. However, the financial system has been effectively privatised because almost all money comes into existence as bank deposits when banks lend money to their customers. The banks create the money that they lend and money is backed, not by gold as it used to be a long time ago, but by debt – by promises to repay loans to bankers with interest.

The important point here is that, while the banks create the money that they lend, they do not simultaneously create the money that their customers also need in order to pay the interest on their bank borrowings. The economy has to keep on growing in order for there to be a basis to motivate new lending. Without new lending, and hence new debt being created, there is no source for the next round of additional money needed to pay the interest on the previous debt.

This kind of economy does not have a reverse gear. Because of its debt based money arrangements the economy must keep growing or the banks get into trouble. If the economy does not grow then it needs something else to grow instead – like asset value bubbles, mainly in the real estate markets, so that banks have a basis to keep on inflating their lending.

If the banks stop lending and people repay their debts the money supply and liquidity starts to dry up, people get into trouble repaying debts and the banks get into problems too. Remember – since almost all money is backed by debt, then in those times when the main dynamic in the economy is that debts are being repaid the money in circulation starts to fall and demand starts to shrink. Horror of horrors the process becomes a vicious deflationary cycle. The economy goes into a downward spiral. Confidence in the banks begins to wobble and people want to take the money out of the machines in the wall.

This explains at a deeper structural level why growth is the taken for granted and self evident goal that few politicians, economists or journalists dare question. It enables us to understand the toxic group-think of the political economic elite – I write 'toxic' because, as argued earlier, it is impossible to reduce carbon emissions sufficiently if the economy does keep on growing.... which means, conversely that, at least when growth does stall, so too do carbon emissions....

...it also gives us an important topic for dialogue with the movement for deep change that has suddenly emerged in tents in cities all over the world, a movement focused on seeking to challenge a crisis of injustice whose roots are the banking system.

Policy Making as in a club...of addicts

So problem number one is that, if you argue the case for policies that would cut emissions adequately, you will be arguing for the ultimate heresy – no growth economics and you will get ignored by the growth junkies – at the same time however we have something important to say to the movement in the street.

That's not all either. Most kinds of addicts share their lifestyle with others – it is so hard to give up their addiction not only because of a brain-chemical dependency but because it means giving up on a social network. That's partly why groups like Alcoholics Anonymous work so well – AA gives another social network based on staying off the drink.

Money and energy junkies are not that different. Inside the addiction circle of very important people it is difficult to get a look in for other ideas anyway. Policy is largely formulated by officials in a dialogue with vested interests or 'stakeholders'. Some lobbyists are much more influential than others. These are the ones well connected to people who own newspapers or other mass media and the journalists working for them. To a large extent public relation companies set the agenda. People of influence have been to the same public schools as the politicians, meet regularly in the same clubs, set up their own think tanks, set up foundations to fund pet causes and operate both behind-the-scenes – or in front of the cameras – all in a way that people without money and time are unable to do. It is in this way that the 1% consolidate their position in the corridors of power.

Regulatory Capture

This helps explain what is called "regulatory capture". The officials working for ministries and public departments which are supposed to regulate private interests instead develop a cosy relationship with those same interests. It seems quite natural for people in a particular economic sector, who have some knowledge of it, to apply for jobs in the regulatory agencies. Likewise people in government, and in the regulatory agencies, regularly take jobs in the very same sectors which they previously had a role in regulating. It is true in the banking sector, which, as an increasing number of people are aware, has taken over and neutralised state regulation. It is also, to a very large degree, true in the energy sector.

Nobody likes to maintain stressful confrontational relationships with others over long periods. It is more congenial when relationships between regulators and regulated are cosy. Then poachers and gamekeepers can switch roles from time to time too. People outside the comfortable clubs, who are losing out, may try to rock the boat to get a problem dealt with – but will often need considerable resources and endurance to maintain pressure to get anything done, particularly if it involves bad vibes.

If they have that endurance, the resources and a good case, outsiders like critical NGOs may, in some cases, be an embarrassment – so they may then be co-opted. Concessions may be made and the critics are allowed to join the club and become instead a force for inertia. Their radical rhetoric gives the appearance that the democratic and consultative system is working.

In the relationship between governments and commercial interests there are few businesses more powerful than the fossil energy companies and the industries closely connected to them – e.g pharmaceuticals. Wherever one looks in the world fossil energy companies and states exist in a symbiotic relationship. Political economic power goes with the deployment of technologies, infrastructures and armaments that use huge quantities of fossil energy. The companies that deliver that energy are therefore of strategic importance and are tightly bound into governments. It is not exaggerating too much to say that either energy companies own the state or, in some cases, are owned by the state. A revolving door relationship exists at the highest level between the personnel of the energy companies and those of governments. What's more, support for democracy takes second place when it comes to securing fossil energy – one has only to point to the cosy relationship between western governments and the autocrats in oil producing countries like Saudi Arabia. Perhaps only the banks have more influence than the energy corporations.

It is against this huge inertia that climate policy in general, and cap and share in particular, have to be developed. The capacity of the political system and vested interests to fundamentally reform themselves is very limited.

On first impressions, given this context, the situation appears to be pretty hopeless. It is certainly an illusion to imagine that a clearly articulated argument about the survival of life on Earth, and social justice, is enough to make a difference in the policy arena as thus described. Even brilliantly expressed arguments can be ignored and they are ignored. One can even define power as 'the ability to ignore'. The higher up the political hierarchy one goes the better at ignoring other ideas and agendas the post holders become. Indeed they have to ignore others because the number of issues that they have to deal with becomes too great. Power holders choose their agendas for focus and ignore the rest. In this regards the whole purpose of seeking power is to pursue one's own agenda choices.

The source of change lies outside the mainstream

However, this is to misunderstand the sources of change, which lie outside the mainstream. The physicist Max Planck, described how change occurs in science – and his words also give us a clue as to how it might possibly change in society and in the economy too:

"An important scientific innovation rarely makes its way rapidly winning over and converting its opponents: it rarely happens that Saul becomes Paul. What does happen is that its opponents gradually die out and that the growing generation is familiarised with the idea from the beginning" [5]

The alternatives to the present log jam have to be constructed outside the political and economic mainstream. Preparations are needed for a rapid transfer over to a new system that is running in embryo when things begin to breakdown, when an older generation flounder and prove quite unable to understand what is going on and quite incapable of coping.

How that might happen has been explored in various writings by different authors and activists who have looked for concept systems that put their local and limited activities in a broader context. The slogan "Think global and act local" is now well known and much recent thought has gone into working out, more exactly, what the phrase, "think global" actually means and how it touches on local practice. In cities all over the world an active movement for change is seeking for how this be done – refusing to prematurely focus on demands, because there is a realisation that this is a complex task and if you are going a long way you need to travel slowly.

Big Ideas and Grand Narratives – for inspirational intrinsic motivations

We are entering a period of great economic and social turmoil and millions of people are showing clearly that they are yearning for a clear way forward out of the chaos. This will be a time when people will be looking for big picture explanations as to what is happening and big picture credible ideas as to the way out. Policies and ideas for climate change mitigation must become an integral part of these big picture narratives. It is important that our ideas are there otherwise the mass movements will be in danger of over-simplifying, thinking that if only we get rid of the bankers then all our problems will be solved.

New arrangements and new thinking about the management of commons is part of the big picture for a future transformation. People are much more prepared to do "their bit" when they feel that what they are doing is part of a larger whole. This gives greater meaning to lives which would otherwise be small because lived in the pursuit of trivial purposes.

In times of turmoil people struggle to understand the bigger picture and embrace new purposes which provide a focus for new intrinsic motivations. As they struggle to understand, to orientate themselves, and to find a way forward that makes sense, they discover causes for themselves in the sense explained by Arnold Bennett. ("A cause may be inconvenient, but it's magnificent. It's like champagne or high heels, and one must be prepared to suffer for it").

This should be compared to the approach of mainstream economics to climate change which proposes that we be nudged towards climate mitigation by changes in prices. It suggests that we need to be "incentivised" – and that we will make money or save money by doing climate mitigation – an approach that relies on extrinsic motivations.

In complete contrast to solving the climate crisis through cash based incentives we need a "Big Idea" which will provide a focus for intrinsic motivations – creating a movement of people working to protect and share common resources. That "Big Idea" is a programme for commons management arising out of a convergence of thinking from different places and requiring new structures and processes based on collaborative networks.

It is beginning to happen. In the years and months before the camps of tents people and movements who have seen their role as protecting common natural resources (the oceans, the atmosphere, fresh water resources) are coming together in a dialogue with those who have realised that they are creating and seeking to defend an "information commons". In the internet, resources like linux, wikipedia and other design processes are effectively creating resources for free in peer to peer work relationships – which corporations try to recapture and enclose to privatise the value created by others for themselves.

"The commons" is therefore a key "big idea" for a policy and ideological platform. **Cap and share**, as well as our other climate policy approaches, needs to be clearly contextualised as an approach that fits best with the management of the earth's atmosphere and climate as a common resource, co-managed and shared by all for the benefit of all, including future generations.

It should be acknowledged here that there is a point of view that big idea explanations which a signpost the future, 'grand narratives' as they are called, are unwanted and dangerous. The fear is that signposting "inevitable and necessary futures" for billions of people to march towards would bring new tyrannies into being. This implies that if we develop global policies like **cap and share** to deal with global problems then inevitably we need global bureaucratic hierarchies with immense power. The worry is that these bureaucracies will, in turn, morph into new top down regimes. Such tyrannies will have great power because, in the face of the big picture which underpins them ideologically, the grandeur and importance of the end – preventing runaway climate change and ecological collapse – would justify any means.

But the point here is that cap and share does not require a big bureaucracy. It is appealing because of its very simplicity.

It is certainly true that a great deal would need to be done to adjust to, and cope with, a rapidly tightening cap. A lot of things will have to be done at the level of households, communities, and in each locality. In each case there will be a need for a unique and location-specific transformation of energy technologies, buildings, production systems, as well as cultivational landscapes and transport configurations.

The Great Transition

Another way of thinking about the tasks at hand is to use the ideas of the "Great Transition". [6] The authors and activists who are developing this overarching framework describe the economy and society as existing in three zones or spheres: a cultural landscape, the dominant economic and political regime and a realm of niche alternatives.

The "cultural landscape" consists of the common motivations of the people in a society and the narratives that the people use to understand the way the world works and their place in it – this is dominant culture of that society.

The culture is embedded and embodied in a second sphere – the economic and political "regimes". These are the powerful institutions that take decisions and allocate resources that have already been described as tightly integrated together. Most of the effort of NGOs and civil society organisations is currently focused on trying to influence these regimes – but as we have seen – these efforts are often too weak and are frequently ignored – or where they do have an impact they tend to be co-opted and then neutralised.

Nevertheless a third zone does exist as a potential source of change. It is not currently very large but it can be found as a place of niche experiments, of small scale developmental projects. In the theory of the "Great Transition" these niche experiments are described as "seed projects". They are run on motivations and narratives which do not fit into the cultural mainstream. If you talk to people in Transition Initiatives, in community gardens, or urban farms, or community energy projects you will typically find that they share a similar story about how the future is likely to look, or how they would like it to look. You will find too that they are much more committed to helping vulnerable people and not in it primarily to make money for themselves. [7]

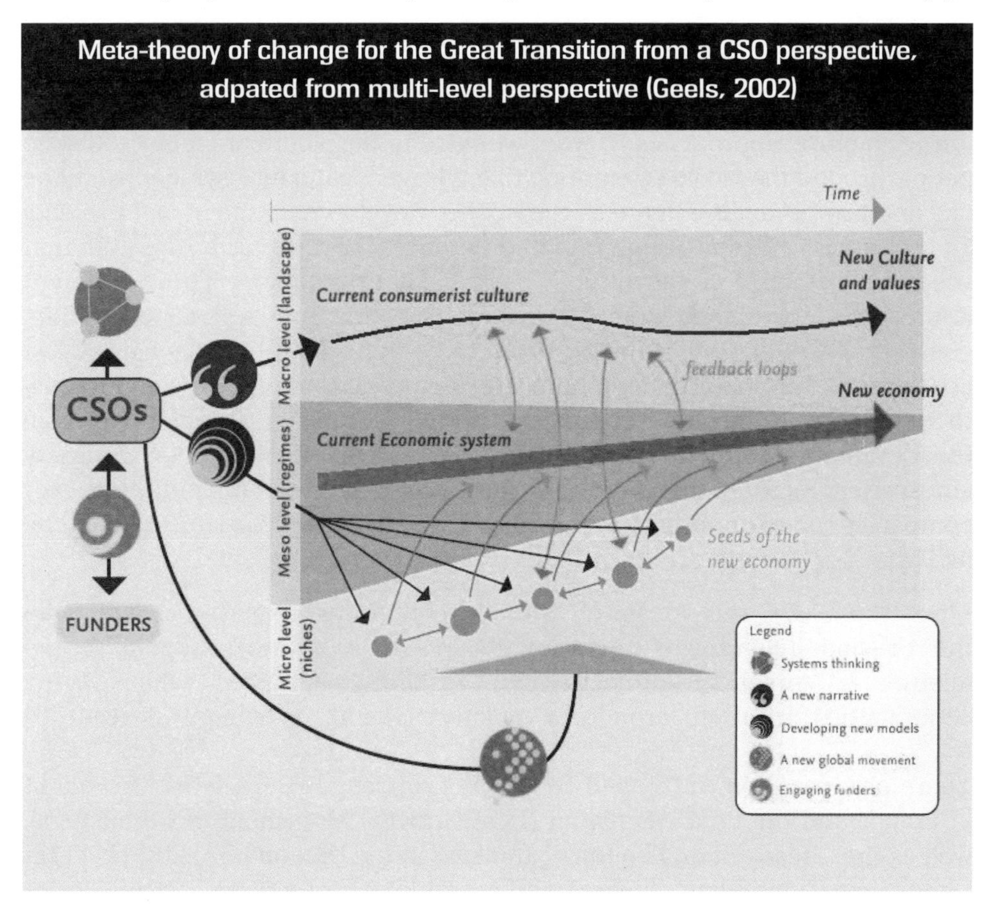

Meta-theory of change for the Great Transition from a CSO perspective, adpated from multi-level perspective (Geels, 2002)

"Seed projects" like this are embodied and embedded in different motivations and narratives and represent an alternative culture. It is by these projects developing further, networking together, becoming stronger that they might start to look more credible as the embryonic basis of new regime, a new economy that is more appropriate to the troubled times.

Since climate change is a system problem, rooted in the use of fossil energy to power a "consume more – bigger – faster" economy, an economy that is getting more unequal all the time, it follows that a lot more than a single policy will be required to get to deal with it. A system change is needed – but how does one achieve system change? It is one thing to explain how the current system works, it is quite another to explain how a transition is to be made from this system to another one that is non destructive. Agreeing what the transition will look like, and then making it happen, in a collective social process is the task at hand. And it is a very challenging task. Nor are we likely to find many mainstream businesses lending a hand as they have an inbuilt growth imperative. This is a task for people – not people playing roles in institutions.

Variants of the Great Transition

The commons and the Great Transition are just two of a growing number of 'bigger picture' approaches to what we have to do. There are many different people around the world attempting to envisage what a new system would be like and how it might emerge as the further development of current projects and practices. Despite superficial differences of approach the different eco-social "visions of the future" have a lot in common. These different approaches are mostly compatible and can converge with one another.

For example, from Germany there are ideas being developed about how to develop a "solidarity economy" – which is forseen as emerging from the development and networking of co-operatives, social and community enterprises focused on ecological and energy goals, community energy companies, community gardens, community supported agriculture and the like.[8]

Then there are the ideas of the Transition Movement, originated in the UK and Ireland and now globally spread – a town and city based mixture of practical projects and reskilling activities which bring communities together around a positive vision of energy descent.[9]

There are ideas too developed by the Decroissance (Degrowth) Movement in France and the Post Wachstum (Post Growth) Movement in Germany, as well as the "Steady State Economy" thinkers in the UK and USA. [10] [11] [12] Under these 'umbrellas' thinkers and practitioners of alternative economics have come together. They seek to counterpose different lifestyles, economic arrangements and the projects associated with them to the growth fetishism in the economic and cultural mainstream. Major conferences on the theme of post growth economics have occurred in Barcelona, Paris and Berlin over the last few years. The conference in Berlin in May 2011 was attended by 2,500 people, many of them young. This is an up and coming generation who will make the future. When we write about the reduction of carbon emissions up to 2050 we are talking about the bulk of their working lives. What they think and the ideas they share will be the future "Zeitgeist".

As is obvious these movements are best visualised as overlapping networks of smaller groups, either of campaigners and/or of project activities – whose ideas are mostly either very similar, or at least mutually compatible and non competitive.

Thus what seems to be emerging is not an "alternative system" like "socialism" and/or "communism" was envisaged to be – a centralised, pre-conceived system created and driven through top down hierarchical relationships and established through a violent seizure of state power. Rather this is a gradual 'bottom upwards' process of local level projects and groups that are networking and cross fertilising in a variety of local, national and international forums. Nor is this an internationalisation that is merely between the rich countries. There has been participation by groups and communities from countries of the south in these discussions which have been more than formal and tokenistic. For example, the ideas of 'Buen Vivir' from indigenous communities in Bolivia and Ecuador, pre-colonial ideas of a good life with harmony between peoples and with nature, have been important in shaping the understanding how to motivate and guide a good life beyond growth.[13]

In his book "Sacred Unrest" Paul Hawken suggests that there are perhaps a half million organisations around the world focused on local economic development, environmental and social justice, and the rights of indigenous people.[14] The characteristics of all these organisations are their huge diversity. Yet they can evolve together in a coherent way. As Elinor Ostrom has argued, it will require a poly centric and multi level approach if the variety of each group, appropriate to the place it operates, is to be recognised and continued and yet, at the same time, the very different groups are to operate together coherently. [15] What is required here is not a top-down bureaucracy which would be incapable of coping with the variety. We need, instead, a network of active groups in which each node has a high level of autonomy and in which overall coherence is created by mutually adaptive arrangements, activated horizontally only as and when needed. Such mutual adaptation between autonomous nodes of activity would be organised to minimise conflicts, maximise synergy, to create and share information about evolving operating environments and to share and create joint value systems to facilitate the common sense of purpose. The ideas here parallel those of the late Stafford Beer and his Viable Systems Model, with its approach to networked management and nested organisations. [16]

This kind of networked coming together is the best hope for the world. It is out of this coming together that we can see how climate action and policy can be shaped in the future. In the meta systems that then emerge to link local level responses we will find ourselves creating a new operating environment in which firstly local government and then national governments will find themselves operating – having a powerful influence on their policy making process. The only sustainable and resilient world economy, an economy which is not climate destructive, is based on a re-localisation of economies. So it is through the networking of local initiatives to grow a greater power that a coherent response to climate change can be developed.[17]

It is processes like these which make it credible to believe that in the future a Climate Trust of the type envisaged by John Jopling in his chapter will emerge. (See chapter 5)

This, in turn, might then provide a different kind of future as the carbon intensive regime of the old men looks less and less credible and has more and more trouble sustaining itself.

Generation change and Zeitgeist

There are good reasons to believe that we are entering a period in which the economy developed and managed by baby boomers on the brink of retirement will find sustaining itself very difficult. Much has been written about the impending peak of world oil production and a peak in world gas production to soon follow. Many argue that we are already at the oil peak and, in fact, on a plateau, so that the decline is soon to come. What is less well known is that this is occurring at a time of generation change in the oil and gas industry itself.

Almost exactly at the time predicted by the IEA for an energy crunch, there is a retirement peak in the oil and gas industry – and this is an international phenomenon. At the time of writing roughly half of the overall professional workforce in production and exploration are aged between 40 and 50 while barely 15% are in their early 20s to mid 30s. According to Booz Allen Research about 33% of those employed in the industry will retire by 2012. It is against this background that we should assess the oil and gas industries ability to rise to the technological challenges like that of successfully and safely tapping of ultra deep water oil. [18] [19]

A similar phenomena can be found in the nuclear industry. There have been serious problems, delays and major cost overruns in the attempts to build a nuclear reactor in Finland. An article in *Der Spiegel* drew attention to no less than 3,000 construction faults at Olkiluoto. In large part this is because the expertise is not there and this problem is due to get worse. 40% of the personnel in US nuclear plants are due to retire soon and the industry will have to recruit 26,000 over the next decade even if it does not build a single new reactor. In 2008 however US universities turned out 841 graduates. The situation in Germany is even more alarming where between 1998 and 2002 only two students graduated in nuclear engineering prompting Areva to fund a training facility in Karlsruhe.[20]

At the same time any young person with an ability to read who is interested in technology and engineering and who is starting their careers are bound to have noticed that “green jobs” are being touted as much more recession proof and also that employment in “green jobs” – or in “cleantech” – are growing fast – albeit starting from a very low point. The idea that these jobs are more in tune with the future is a plausible one because, while the fossil fuel and nuclear sectors are running down with their engineers and their personnel are retiring, over the very same period “clean energy” employment is growing rapidly. This is especially the case in countries like Germany and China but it is even the case in the USA where employment in this sector grew by 9.1% per annum between 1998 and 2007. [21]

The point is that employment changes in terms of training and new job entry at one pole and retirement at the other pole are lagging indicators of a trend that is occurring now anyway. The twin process can be seen as a powerful reinforcing feedback in a transition that is and will occur consisting of an acceleration of the decline of traditional carbon based energy sectors and creating an upward dynamic in those replacing them.

A contested and confusing transition

Notwithstanding, caution is needed. The employment and generational transition is not occurring, and will not occur, without conflicts and considerable contestation. There is obviously a battle opening up around what the future energy system will be like – gas, nuclear or renewables. These are alternatives and it is not easy for governments to have a mixture.

The crisis at the nuclear reactor at Fukushima after the earthquake and the Tsunami, a crisis which will clearly not go away and which will run for months and months, perhaps year after year, has been a serious blow to the nuclear sector. It has had its greatest effect in Germany where the country appears to have decided to stake its future much more on renewable energy. At the time of writing Germany is looking at how it will upgrade and change its grid to make this possible – perhaps by adapting and upgrading the electric power lines of the railways, thus minimising the nimby backlash. [22]

Simultaneously the oil and gas industry are contesting moves towards a future based on renewable energy. They want political backing for the further development of fossil fuels and, in particular, supporting for so called "unconventional gas" – by technologies which drill into and shatter shale rock formations to release the gas trapped in them.

In the USA shale gas development has become hugely controversial. There are environmental and health effects from the toxic materials that have been used and released into surrounding rocks, water, the atmosphere and soils. In the UK shale gas has been associated with an earthquake near Blackpool. Shale gas is controversial too because the fracking production process, as well as pumping gas from production source to its place of combustion, has been found to entail significant leakage. The natural gas thus leaked, mainly methane, is itself a powerful source of global warming. These facts are undermining the claim that natural gas a source of relatively climate friendly energy. Fracking has been banned in France – but it looks as if it will go ahead in the UK.[23] [24] [25]

The change in energy system is thus contested and its outcome unclear. Nevertheless the fall in oil production after peak is likely to be fast and we are witnessing processes that will progressively change the conditions in which all governments operate. If governments fail to recognise what is going on at this point in time it is because they are still operating under the influence of old men and old financial institutions. The new networks of groups that were described earlier are not strong enough to impose their ideas and will on the state. Will this change soon?

The political system – waking up a bit late to impending chaos

What we have been witnessing are the thrashing agonies of a dying energy system that is using its traditional links and grip on the political system to try to maintain its influence. This influence is, however, beginning to wane. The death agony is well covered up by PR and spin but it is a death agony without doubt.

The political establishment and vested interests have been very resistant to change. As a result we are entering a period of crisis with a woefully unprepared political system. Crises like this are periods of danger but they are also periods of opportunity – because it becomes clear to thinking people that things cannot go on in the same way. It is the preparation for such a generalised crisis that we must now apply ourselves to.

As is generally recognised, the full force of the climate crisis lies some way in the future. However, if not enough is done in the next few years then, by the time the terrifyingly destructive impacts are felt it will be too late to do anything meaningful. The effects will keep on rolling relentlessly for centuries. Nevertheless, here and now, the energy and economic system is about to enter a period of convulsion anyway. *Cap and share* and our climate policies thus need to be made fit for purpose as part of a package that millions of people identify with as being necessary to deal with the structural problems, not in the future but now.

An immediate future of great uncertainty

There are arguments that what we can still do will not be enough when measured against the huge necessities for change required for substantive climate mitigation. This is the argument of Clive Hamilton in his widely praised the book *Requiem for a Species*. However Hamilton assumes that the recession unleashed by the credit crisis which has stabilised global emissions is merely a temporary problem.

This is very unlikely to be the case. As argued it is very probable that we are now in a period of economic instability because of peak oil and peak debt which will continue. Although emissions bounced backed strikingly in 2010 after the recession, one must wonder how long the "recovery" will continue.

In important respects, the instability will not help. Worse case scenarios suggest that the interaction between declining oil supplies and the fragile financial system could cause huge dislocations and these, in turn, could undermine the basis for large scale engineering solutions to energy shortages and the carbon crisis. Under these worse case scenarios the deflationary collapse of the economic system, which at the time of writing seems very likely, would lead to a disintegration of the very fabric of complex economic organisation needed to deliver the components for a renewable based rebuild of the energy infrastructure.

Nevertheless one can turn the pessimistic argument on its head. In the face of floundering economic, industrial and ecological policy in the next few years the best thing to help would be to unify and mobilise all of society behind a major investment programme for energy and agricultural transformation – before it is too late. When societies are in chaos, malevolent elites pick a fight with neighbouring countries and an external enemy creates internal cohesion. An elite that finally realises it must fight to prevent a breakdown of the energy system instead of an external enemy might be able to pull things round. Alternatively this idea of a global fight to renew the energy and the cultivation systems, particularly in a way that stresses commons can provide a large part of the unifying vision for the movements of the streets, offering work and justice at the same time. Once underway, the accumulation of renewable energy equipment and its infrastructure would create its own self feeding dynamic, delivering more energy than it costs to build up. In that kind of future context there is some kind of vision of hope against mass destitution which a collapsing finance sector is bringing down on our heads.

There do seem to be huge opportunities for renewable energy systems – in particular offshore wind energy around the UK and concentrated solar power in southern countries and deserts. There are also opportunities for considerable reductions in energy consumption. There are arguments that, for example, the energy return on energy invested in offshore wind are considerable and the scale of the engineering challenge is no greater than the previous construction of an offshore oil infrastructure.[26]

The open question – Chaos or Grand Transition?

It will be challenging. In some parts of the political system a few officials and politicians are just beginning to get a belated understanding of this. Although there is a great reluctance to transform the energy economy in face of climate change there is the first dawning of a recognition that the energy economy *will have to be* transformed because of peak oil. The code words for 'peak oil' in business and government are 'energy security'. Some parts of the business establishment too has finally acknowledged the message of peak oil and are looking at what will be done about it. Although the peak oil and climate imperatives are not identical they do overlap.

With this growing awareness the danger is that politicians and business will take the wrong decisions. The peaking of conventional oil could worsen climate change by driving an increased use of more carbon intensive substitutes and biomass." In order to keep global temperatures within 2°C or preindustrial levels, cumulative CO_2 emissions must be kept well below the amount per would be produced from burning the remaining proven economically recoverable fossil fuel reserves".

Nevertheless, there is an increasing recognition that if the energy system must be transformed it makes more sense to deal with climate change and peak oil at the same time. Some can see already that it is a dead-end to try to use the remaining fossil fuels and that it makes more sense to go directly over to renewables.

An example is the Centre for Alternative Technology's second addition of Zero Carbon Britain – *Zero Carbon Britain 2030* – in which cap and share is described as one of a number of possible policies in the framework that will be needed to drive decarbonisation. [27] Another example of cap and share in a general package of policies, from our own ranks is the Holyrood 350 Programme for Scotland. [28]

Another example of a policy which connects action on energy security (peak oil) with action on climate change is a Lloyds/Chatham house report on sustainable energy security. This argues that:

"Energy security is now inseparable from the transition to a low carbon economy and business plans should prepare for this new reality. Security of supply and emissions reductions objectives should be addressed equally as prioritising one over the other will increase the risk of stranded investments or requirements for expensive retrofitting." [29]

In summary we can expect to see energy transformation being pushed up the political agenda. In the best scenarios we would expect to see a search for new and more effective, policy mechanisms for carbon reduction occuring too. This is because, while it is becoming blindingly obvious that these are absolutely core issues, the global political establishment has clearly shunted itself into a dead end in trying to do something about these issues.

The Copenhagen Debacle

In this regard there is a most extraordinary situation opening up. For all the reasons explained at the beginning of this chapter the global political and economic elite have totally failed to provide anything at all credible in the way of a response to the climate crisis and energy crisis. The collapse of the UNFCCC process at Copenhagen and the collapse of Obama's efforts to introduce climate legislation in the USA can be seen as a stalemate between an old energy order and political system and a new one that is not yet powerful enough to emerge and make its dynamic the dominating one. The energy system of the old men and old money is still too powerful. But, as we have seen most of these old men will have gone in a very few years – and the carbon energy that they supply, and which is their power basis will be in precipitate decline.

We are, in short, moving towards a situation where policies like *cap and share* and a *carbon maintenance fund* to prevent loss of soil carbon need to be argued for as part of packages of transformation in order to avert a generalised collapse caused by the wooden headedness of fixated old men. The support for the new developments will largely have to be found outside the political mainstream in the emerging new movements that were mentioned earlier in this chapter.

Through the projects and networks of these movements only so much can be done in energy efficiency and carbon reduction at the household community and local level if there is not some wider framework to "lock in" what is achieved. Without an adequate framework the improvements that are made would be immediately lost because of "rebound effects" of the type explored by Nick Bardsley in his chapter. Also, the energy and carbon saved in one place would be squandered by irresponsible people and companies in another place. These community-based activities will inevitably be driven by a stronger prioritisation for social justice issues and the share in cap and share will be more attractive and influential here.

The key idea here, to return to the idea of indirectness, is that climate policies need to be not just head on attempts to tackle climate change but ideas for society – for reconstructing energy systems, for maintaining macro economic activity and employment (if not growth), for expressing new ideas of social justice and also for making clear how we are going to look after each other. Let us now turn to these points.

The macro economics of climate policy and the politics of rent at the limits to growth

Given the wider picture we should not forget that cap and share can be promoted not only for driving decarbonisation but because of its effect on purchasing power as energy prices rise. Cap and share has more to offer than as a driver of climate mitigation alone. After peak oil each new impetus to economic recovery is likely to lead to a spike in oil prices that will, in turn, crash the world economy. This volatility will not help long run structural changes and nor give the security needed to encourage productive investment in new energy systems.

As fossil energy prices soar upwards many non-marginal energy producers will, for example, still be supplying from fields with low production costs. During the price spikes they will be raking in money way above their production costs and there will be a transfer of what economists call "scarcity rents" to these producers. (Rent is here the large amount of money made when there are high prices because of high demand and scarcity even though some producers are still able to pump oil and gas relatively cheaply). These "rents" will be taken from the pockets of everyone else. Rent transfers like this unbalance the economy, lead to unrest and bring on the next crash.

Beyond the "limits to growth" there is still room for money junkies to get rich if we let our unjust system continue – not because of their inventiveness, or their enterprise, or what they produce, but because they succeed in cornering the ownership of the scarce resources that everyone needs – energy, the atmosphere, fresh water, land, food commodities... and then are able to charge a high price, enriching themselves while the poor are driven into destitution. This threatens to be a 21st-century "politics of rent" and we have to find answers to it.

From this point of view, arrangements like cap and share have a wider relevance. It is necessary to manage the Earth's atmosphere as a global commons for which we are all equally responsible, in a way that ensures that, when there are benefits to be had, we all get them – also ensuring that particular groups are not unfairly burdened. The energy transformation should be arranged in such a way as to ensure that the mass of the global population get a share from the sale of permits. This will balance purchasing power, moderate the contractionary process and part-provide some of the capital resources needed to help people transform their homes and gardens. At the same time it can help provide the incentives and stability for large investments, like offshore wind, where there are the resources, the capacity and will for these to be developed.

Similar principles need to be applied, adapted to context, across other natural and human commons – including in the monetary system regarded as a Commons. Earlier we explained that the debt based money system is a major part of the problem. It has no reverse gear and is implicated in the growth fixation of mainstream politics. That's because the money commons has been privatised in the interests of the moneylenders and a major part of the overhaul that is needed is to transform the money system too – to manage it in the interests of everyone.

Commons resources should be managed in the interests of all – including future generations who should inherit them intact and healthy – the oceans, fresh water supplies, land rent and the like. That means not only policies but appropriate institutional architectures. These are major agenda items but it seems unlikely that top down policies from governments will emerge until a lot of bottom upwards improvisation from grass roots movements of the type described earlier has been tried and been found to be practical and workable. [30]

Conclusion

To sum up, the existing economic and political system has proved incapable up to now of embracing anything like an adequate level of climate mitigation. It can be argued plausibly that it is already too late to prevent runaway climate change. It is certainly touch and go. There is nothing inevitable about the future. Nevertheless it is clear that we are entering a period of economic, social and political turmoil brought about by peak oil, peak debt and the decomposition of a political system that millions of people now regard as corrupt and not to be trusted. There is increasing recognition even in parts of the business elite that major changes in the transformation of the energy system are going to be needed and that it does not make sense to deal separately with peak oil and climate change. In this context the relevance of policy ideas like cap and share to these other problems must be made clear and such policies firmly located in packages for transformation.

What is still not a clear is how far governments are capable of contributing to the new future. There is an argument that states are being increasingly hollowed out and incapable of real social and ecological leadership. It has been argued, for example by Naomi Klein, that states are led by parties functioning as brands, backed by PR machines, intent on organising society to whatever the financiers want. It is certainly this view that makes most sense of the utter failure of the state to control the financial markets and the financial sector. [31]

Of course, we want government support for what we are doing if they will give it – but meanwhile if there is indifference and hostility from governments then we must get on and set up the organisations that we need. We can do that by setting up organisations where we are and then networking them together. When we do this we do it in the hope that there will be a supportive government buy-in later, when pressured by our movements with their different commons based ideologies and their practical community relevance on the ground. If we cannot get governments to do the job we must move to set up the organisations that we need and then struggle to win them the power to do the job directly. At the current time governments will not go against the elite consensus – but in the profound turmoil ahead we should not underestimate the extent to which power relationships will change if we are well organised, with clear ideas that attract a mass following.

In this context it makes sense to evolve a package of economic energy and climate policies to address the different crises together – financial, energy, climate, cultivational. Such packages of policies which seek to reconfigure the world we live in have already begun to appear – like "Zero Carbon Britain 2030" and the programme of "Holyrood 350" in Scotland. These programmes will be immeasurably strengthened by being based in new forms of networked commons organisations operating with charters of rights and responsibilities that they win from the existing political system.

To fully complete the reconfiguration of our economy and society we need to connect with the emerging movements with new ideas that captures rights to defend the commons and new ways of managing them which do not rely on yet another top-heavy bureaucracy .

Endnotes

1. John Kay, *"Obliquity"* Profile Books, 2010.
2. Tim Jackson, *Prosperity without Growth?* Sustainable Development Commission, 2009
3. Roy Madron and John Jopling, *Gaian Democracies*, Schumacher Briefings/Green Books, 2003
4. Clive Hamilton *"Requiem for a Species"* Earthscan, 2010, p33.
5. Max Planck *The Philosophy Of Physics*, W. W. Norton & Co. 1963
6. http://www.smart-csos.org/
7. See for example, Nadia Johanisova, *"Living in the Cracks"*, Feasta/Green Books, 2005
8. www.solidarische-oekonomie.de
9. http://www.transitionnetwork.org/
10. http://www.decroissance.org/
11. http://www.jenseits-des-wachstums.de/
12. http://steadystate.org/

13. Thomas Fatheuer, *Buen Vivir*, Hsg. Heinrich Boell Stiftung, Band 17, 2011
14. Paul Hawken, *Blessed Unrest*, Penguin Books, 2007
15. Elinor Ostrom, *A Multiscale Approach to Coping with Climate Change and other collective action problems*, http://www.thesolutionsjournal.com/node/565
16. Stafford Beer, *Think before you Think*, Wavestone Press, 2009, pp134-157. See also Jon Walker's article at http://www.esrad.org.uk/resources/vsmg_3/screen.php?page=preface
17. Richard Douthwaite, *Short Circuit: Strengthening Local Economies in an Unstable World*, online edition, June 2003 downloadable at http://www.feasta.org/2003/06/16/short-circuit/
18. Jodie Humphries, *Oil and gas workforce – a shortage in skilled labour*, Jodie Humphries August 2010 at www.ngoilgasmena.com/article/oil-and-gas-workforce-a-shortage-in-skilled-labour/
19. http://www.cres.ch/Documents/SKILLS%20SHORTAGE%20PART%20I%20pdf.pdf
20. http://www.spiegel.de/international/europe/0,1518,655409-2,00.html
21. http://www.enn.com/top_stories/article/40051
22. "*Germany explores using Train Lines as a Power Grid*" http://www.spiegel.de/international/germany/0,1518,758698,00.html
23. http://news.bbc.co.uk/1/hi/programmes/newsnight/9255520.stm
24. http://www.nytimes.com/2011/03/04/us/04gas.html?_r=2&hp
25. Robert W Howard, Renee Santoro, Antony Ingraffea, *Methane and the greenhouse gas footprint of natural gas from shale formations*. A letter. Climatic Change, Accepted March 2011
26. http://graphics8.nytimes.com/images/blogs/greeninc/Howarth2011.pdf
27. The Offshore Valuation Group, *A Valuation of the UK's offshore renewable energy resource, published by the Public Interest Resource Centre*, 2010
28. http://www.zerocarbonbritain.com/
29. http://holyrood350.org/
30. Lloyds/Chatham House Report "*White Paper. Sustainable energy security. Strategic risks and opportunities for business*" www.chathamhouse.org.uk/files/16720_0610_froggatt_lahn.pdf
31. http://www.boell.de/economysocial/economy/economy-commons-10451.html
32. www.alternet.org/media/145218/naomi_klein:_how_corporate_branding_took_over_the_white_house?page=entire

Chapter 2

The Climate and the Commons

Justin Kenrick

This chapter aims to highlight the continued existence of the 'Commons', a viable alternative to the socio-economic system which is driving climate change. It explores strategies for supporting and restoring the commons at the local, national and global levels.

The premise of this chapter is that there is a proliferating movement of initiatives seeking to defend the commons (mostly in the Global South) or restore the commons (mostly in the Global North), to ensure our survival and well-being. This chapter is also premised on the notion that we still have time to act to restore our socio-ecological sustainability. Adding together the temperature rise industrial society has already caused since the start of industrialisation – 0.8ºC – and the unavoidable temperature rise in the pipeline from the last 30 years of emissions -: 0.6ºC, we arrive at a likely rise of around 1.4ºC. The first point of no return where tipping points could become irreversible is estimated to be somewhere approaching 2ºC.

Kevin Anderson (of the Tyndall Centre for Climate Change Research) has argued that current international negotiations, even if successful, would be likely to take us to 4ºC [1]. To hold at 3ºC (if hold is possible at that temperature, with arctic melt, forest fires, and methane release feedbacks underway) would – he argued – require a 9% cut in emissions per year. Anderson argued that the only way we know to cut emissions rapidly is economic recession: the collapse of the Soviet economy cut emissions there by 5% a year, half the rate of reduction needed now [2].

So to not pass the 2ºC tipping point we clearly need to stop the industrial economic growth machine in its tracks, farm in a way that improves the fertility of our soils, and enable the regeneration of our forests. This is clearly completely unrealistic and impossible from within the mainstream mind-set and policy options. The only way we can do this is if we understand, defend and recreate ways of meeting our needs which ensure sufficiency for all without devastating societies and ecosystems. Such commons systems have always persisted beyond, within and despite the dominant economic system, so what are they?

1. Introduction: Holding the land in common

The morning mist obscures the mountain, and smoke rises through the thatched roofs of the round or oblong earth and wood walled huts, as people, goats, sheep and cows wake...

Cows munch on the nearby grass, their bells tinkling, goats call from their little huts on stilts where they have slept overnight, defecating through the slatted wooden floor, and sheep emerge from the room they have slept in at the end of the oblong houses. There are two other rooms in these homes: one for the family, and the other a large room that acts as kitchen and living room with its firepit in the middle of the earth floor. There were ten of us sleeping on that floor after talking, music and laughter late into the night.

This is one of the villages where the Ogiek live in their ancestral homeland of Chepkitale on the slopes of Mount Elgon, Kenya. We are here to assess whether their ways of living are compatible with conservation's stated aim of conserving the high moorland they live in, and conserving the forests on the lower slopes that they use for their cattle in the dry season, for honey and for gathering fallen wood. I am working for a human rights NGO that the Ogiek have asked to help stop them from being evicted. We have brought a team of conservationists and local councillors to evaluate their way of living. It quickly becomes clear that they are living in a way that is a million times more sustainable than any of us who are evaluating them. Their knowledge of medicinal herbs is profound, their answers to sceptical conservationists shows a depth of understanding of their environment born out of generations of ensuring that the forest and wildlife and livestock are all kept in balance, since this is the source of their well-being. In such a system, it is clear that there is no contradiction between self-interest and inter-species altruism.

The clearest evidence for the interdependence of the whole system comes from accounts of what has happened when the Ogiek have been forced off their land by a combination of conservation-logic ('nature is better off without people') and development-logic ('people's lives improve by being embedded in a competitive cash economy'). The result has been that poachers have used the opportunity of their absence to decimate the elephant population, and charcoal burners and illegal loggers have moved in to cut the forest. The Ogiek are very clear that their livestock and the wildlife co-exist happily, that their presence deters poachers, and that they will resist illegal and legal loggers who destroy the indigenous forest on which they, their cattle and the bees depend.

All of that was very clear, and yet when we return to the big meetings in Nairobi the crazy logic of enclosure reasserts itself. Some of the officials in Ministries and large conservation and forestry bodies reassert the notion that the Ogiek are living a miserable life and that pushing them out of their land is for their own good. They reassert the notion that to protect the forests, the wildlife, and the water catchment area, the people must go and the area must be taken over by conservationists (and presumably be off-limits to all humans apart from researchers, guards with guns and tourists with dollars). Thinking back to the high spring where one of the key rivers rises in the moorlands, to the trees that protect it, and the hoof prints of antelope and sheep that had drunk at it – it was clear that we were encountering the biggest lie. This is the lie that humans don't belong, and that our only hope is to place our faith in separating ourselves from nature, in controlling and extracting from nature, and in controlling and competing with each other.

The Transition back to Sustainability

This example from Kenya is repeated across the world anywhere people are living in sustainable ways, and has been repeated over the centuries as communities have been forced to make the transition to unsustainability. This is a transition most communities in England, Wales, Ireland and Scotland were forced to undergo generations ago through enclosures, clearances and internal colonisation, a process that was spread to the rest of the world through colonisation and which is still underway. The power of the transition to unsustainability relies on the threat of physical force backed up by the story that there is no alternative: that there is no way of belonging.

Here in Scotland (as I am sure elsewhere) that story has been given the lie over the last 15 years, as communities on the west coast have taken back their land into community ownership. What began with Assynt and the Isle of Eigg reclaiming their community lands from absentee landowners in the nineties, was then taken up by the new Scottish Parliament which – through the 2003 Land Reform Act – provided the framework, and for a while the financial support, to enable communities to reverse the clearances, the enclosures, and to reassert community control of their lands. Over 500,000 acres is now back in community hands, before the funding faltered and the movement slowed, awaiting a resurgence of community and political leadership that is beginning to step forward as the reality of the economic crash hits home (for example, on 20th February 2012 the Scottish Government committed £6 million over 3 years to a new Scottish Land Fund to support community buy outs). We are reminded that real wealth lies in resilient communities not in bank balances that can vanish overnight.

Here in Portobello, Edinburgh, we started our local transition initiative – PEDAL – in 2005 before the word 'Transition' had emerged, inspired by what Rob Hopkins and his students were doing in Kinsale in Ireland. PEDAL stands for 'Portobello Energy Descent and Land Reform Group', and it was clear from the start that land reform is as crucial to making the transition back to sustainability as is the focus on energy descent, on reducing our dependence on oil. In an economy that isn't growing, one person having more than their fair share means someone else going without: and since land is the source of all wealth, sharing this fairly is fundamental to ensuring sufficiency for all.

Like other Transition initiatives we have been finding the gaps and cracks where we can make the difference: finding derelict land we could turn into an orchard, spaces behind shops where we can start community gardens. Like others we have had to negotiate with powerful landlords to try and get agreements for more ambitious projects, such as negotiating with Scottish Water to seek agreement for a site for Portobello and Leith's proposed community turbine. This is all very good, but is it enough?

Land Reform – Should it be Occupying Transition?

As we face the interconnected crisis of ecological destruction (soils, oceans, atmosphere, forests) and resource depletion; and as our system alternates between economic growth that destroys the planet, and economic collapse that destroys livelihoods – surely what is needed is to recover the commons approach that Transition embodies, and to take it to a new level.

What's happening?

On the downside:

Global oil production has been flat since 2005, removing the magic ingredient that made growth seem as though it could last forever. The sale of luxury goods goes through the roof, as the wealthy get wealthier even as the rest of us face cuts and unemployment. The Climate negotiations at Durban failed as we all expected and emissions continue to rise, while orthodox 'solutions' to climate change get slammed even by their most ardent backers. For example, the Swiss Bank UBS's November 2011 report states that: "By 2025, the European Emissions Trading Scheme will have cost consumers 210 billion euros. Had this amount been used in a targeted approach to replace EU's dirtiest plants, emissions could have dropped by 43%, instead of almost zero impact on the back of emissions trading."[3]

On the upside:

North Africa and the Arab Spring, creative protest across Europe, and the Occupy Movement insisting that we find solutions that include us all. Scotland likely to be independent by 2015 and potentially demonstrating a much more community focused response to economic and ecological crises. That community-focused approach may look unlikely now but so much that has happened these last few years has been unforeseen by the so-called political and economic experts – it makes you realise they may be experts in the dominant paradigm but not necessarily in reality.

So, what is the reality we need to be thinking/ making/ being?

Is there a way of fusing Transition's focus on the primary importance of place, and the Occupy movement's focus on the crude fact that the very few are destroying the planet we all depend on? Is there a way of stepping forward and making clear that there is and always has been an alternative, not just at the local level but at the society-wide and global levels? That alternative is evident in the community-focused way humans have always done things until forced into submission by an ideology that believes we don't belong here, and that uses force to make that seem true.

Should we be thinking much more strategically about seeking agreements with councils and corporations but, if that is lacking, then peacefully Occupying the spaces needed for food growing, for energy production, for ensuring the basic needs of care for each other are met? This may seem outlandish to suggest now, but way back in 1940, Priestley suggested that "We must stop thinking in terms of property and power and begin thinking in terms of community and creation"[4]. As the gambling financiers take home their winnings, and as the economy unravels on the back of declining oil supplies, we may well be faced with the kind of choice Italy and Greece have recently faced. And they ultimately had no choice because they had no alternative. Having a clear society-wide alternative to ecology-busting growth and people-smashing austerity (boom and bust) may be one way of making the transition to the next level: branching out without losing the deep roots in place we all depend on.

2. What is the commons? What is its power?

'The commons' refers to resources that are owned in common or shared between or among communities' populations. Globally, between 1 and 2 billion people define their right to the land through what the community says, which means that the commons estate is huge[5].

For example, in sub-Saharan Africa 1.6 billion hectares (or 75% of the land area) is held by communities under customary law. Around 1.4 billion hectares of this is land that is not cultivated but is collectively held forest and rangeland, used to hunt, gather or graze animals on. In most of Africa, private title covers only on average between 2 and 10% of the land. The colonial legacy led to the rest of the land being considered 'without owner', when in fact these vast areas were and are held by communities under common ownership systems. By describing this land as 'without owner', colonial powers justified expropriating it – a situation that has been maintained by many African states after independence[6].

To continue with the example of Africa, the vast majority of forest, pastoral, farming and other land has been managed by communities in accordance with their customary law, which is a dynamic approach to the management of natural resources that continues to develop and respond to modern realities, including issues of population growth and movement, shifting resource use, and food sovereignty[7].

Elinor Ostrom, the winner of the Nobel Prize for Economics for her work on how commons regimes function, outlines why such commons regimes have ensured security for community members and maintained an abundant environment. She writes, "When local users of a forest have a long-term perspective, they are more likely to monitor each other's use of the land, developing rules for behavior. It is an area that standard market theory does not touch."[8]

This standard theory follows Garrett Hardin's famous 'Tragedy of the Commons' parable in which commonly held land is inevitably degraded because everyone in a community is allowed to graze as much livestock as they want to there. But Ostrom's research refutes this abstract concept with real life experience from all over the world, showing that the number of livestock you are allowed to graze in a commons regime is decided by the community as a whole, since although each individual may want to graze as many as possible, the community as a whole knows it needs to ensure the well-being of the environment upon which they all depend[9]. Meanwhile Garrett Hardin himself later revised his own view, noting that what he had described was actually the Tragedy of the Unmanaged Commons, or the Tragedy of Open Access regimes;[10] and one of the best examples of which is capitalism.

Economist Joseph Stiglitz, also a Nobel winner, commented, "Conservatives used the Tragedy of the Commons to argue for [private] property rights, and that efficiency was achieved as people were thrown off the commons... What Ostrom has demonstrated is the existence of social control mechanisms that regulate the use of the commons without having to resort to property rights."[11]

However, it is crucial to be clear that customary commons regimes, such as those that regulate land ownership and use in most of Africa, are based on highly developed and evolving property rights systems which can include the recognition of individual or family rights to certain areas and resources, but always within a larger frame of rights which are community-derived, and which the community adapts to respond to circumstances.

The commons are often mistakenly understood to refer to resources which are intended to be shared as widely as possible, or to refer to resources – such as oceans and the atmosphere – which it may at first sight seem impossible to limit access to. In fact, whether in relation to a stretch of river or coast for fishing, to an area of forest or grazing land, common resource management systems require a very clear *demarcation* of who is included in, and who is excluded from, using the resource. It is on this basis that a system of rights and responsibilities evolves, even if one aspect of such systems tends to be their ability to *flexibly incorporate* new members who are willing to abide by the *reciprocal rules* required to maintain the commons[12].

Ostrom usefully distinguishes 8 design principles underlying successful commons regimes[13]:

- Group **boundaries** are clearly defined.
- *Rules* governing the use of collective goods are well matched to local needs and conditions.
- Most individuals affected by these rules can **participate** in modifying the rules.
- The rights of community members to devise their **own rules** is respected by external authorities.
- A system for monitoring member's behaviour exists; the community members themselves undertake this **monitoring**.
- A graduated system of **sanctions** is used.
- Community members have access to low-cost **conflict resolution** mechanisms.
- For Common Pool Resources that are parts of larger systems: appropriation, provision, monitoring, enforcement, conflict resolution, and governance activities are organized in multiple layers of **nested enterprises**.

Today the term 'the commons' is used to refer to a far broader range of resources that are "held in common", and can include everything from the community commons that is the focus of this chapter, to the cultural commons that animates so many today. The *community commons* can include natural resources and shared community institutions (such as those for resource allocation and dispute settlement, child care and care for the elderly, health care and community provided education) while the *cultural commons* can include literature, music, arts, design, film, video, television, radio, information, open source software and collectively created and maintained internet resources such as Wikipedia.

These two experiences of the commons can run along side each other, be mutually supportive, or run into conflict. For example the community commons tends to focus on – first and foremost – ensuring sufficient resources for all within the resource limits of a particular locality, and sees community ownership as the key to ensuring this sufficiency. In contrast, the cultural commons approach often sees the commons as infinitely abundant. One place where the clash between the two approaches has become apparent has been where well-intentioned northern NGOs have established internet protocols for people across the world to be able to use indigenous knowledge from particular places respectfully, while the communities themselves argue that the use of their knowledge requires not an abstract impersonal protocol but the establishment of real face to face relationships of trust.

From one point of view the government and public services in a democratic society could be considered part of what the members of that society collectively hold in common. From another point of view there is a clear gradation from the libraries and parks and pavements that we collectively use, the elections and jury service through which we can make decisions, the free health and education provision to which we are entitled, to the experience of government, financial institutions and corporations as forces over which we have no control, forces which appropriate the commons rather than protect them. This is nowhere clearer than in corporations legal responsibility being to increase shareholders profit rather than social or environmental well-being; or in the way the UK planning system makes no provision for prioritising community resilience, and always gives the last appeal to external developers who argue that any local disturbance is far outweighed by the public good of economic growth which their money making is seen as inevitably engendering.

The most important point to make about the commons regimes through which a vast number of communities across the world meet their physical, social, cultural and spiritual needs, is that their institutions are not derived from the state or the market but from long-standing experience that people in community can robustly – through discussion and debate, disagreement and consensus – self-organise their affairs; that we do better when we don't have a supposed 'higher authority' to impose its will on us, but when we learn from experience and decide for ourselves.

While between 1 and 2 billion people – especially in the Global South – continue meeting their needs through commons regimes in ways that sustain their environments for future generations, there is a resurgence in commons regimes in the Global North as well.

Commons systems of land ownership and resource use – such as in the forests of Scandinavia or the crofting farming systems in the Highlands of Scotland – have persisted despite the centuries of enclosure during which commons regimes have been forcibly appropriated by the wealthy, their inhabitants being forced to either work for a pittance for the new 'owner' or pushed off their land to search for work in the factories of the cities or to emigrate, often joining the military of Empires which have then been used to take over the land of other communities that have been operating on commons principles elsewhere.

In places like Africa and Australia the colonial expropriation of common lands happened through legally describing them as 'wastelands' or 'empty lands' (terra nullis) and claiming that they were held in property systems that were not defined by private ownership and therefore were not admissible. In the same way, Andy Wightman describes the dispossession of common property regimes in the Highland and Islands of Scotland as being carried out through the power of the state acting on court decisions which were determined by the wealth of the powerful. These were court decisions which the poor could not oppose since – in the title of his most recent book – 'The poor had no lawyers'[14].

As a long delayed reaction to these Highland Clearances, the new Scottish Parliament passed legislation – the Land Reform Act (Scotland) 2003 – which provided legal routes for (and, for a while, financial support to) communities in the Highlands and Islands seeking to bring back their land into common ownership and out of the hands of landowners who had long since dispossessed them. This legislation was in large part a consequence of communities taking the matter into their own hands and finding the collective will and the means to buy back their land (as happened for example in Assynt and the Isle of Eigg), but many other communities have followed suit helped by the legislation (places like the Isle of Gigha, Lewis and South Uist) and such communities now own 500,000 acres of Scotland.

This statutory legal recognition of the (often unwritten) customary ownership systems underpinning commons regimes is critical to the protection and resurgence of commons regimes in the Global North and South.

Just as communities in Scotland who have regained community control of their land, their shared buildings and other resources, are demonstrating an ability to attract people back to their communities and to develop environmentally and socially sound approaches to livelihoods and land management, so in Africa and other parts of the Global South, study after study demonstrates that customary land tenure provides a secure foundation for sustainable development providing that it is protected under statutory law[15].

In places like Africa, the alternative is to continue with an out-dated colonial-derived system that makes communities squatters on their own lands. In this context, the biggest driver of social conflict, and in extreme cases civil war, is the denial of rights to land and resources. We have seen conflict of this kind in recent times in both Central and West Africa where 50% of civil wars resume within 5-10 years when such issues are left unresolved.

A critical role for the modern state – whether in the Global North or South – is to establish ways of recognising and incorporating customary land tenure into national law, recognising that it is fair, adaptable, dynamic and legitimate. Some countries such as Tanzania and Scotland are moving ahead with this, however many countries are far behind and have an urgent need for reform.

The recognition or restitution of customary land tenure provides a route to food sovereignty and a secure basis for development, particularly in the context of large-scale land and asset acquisition by domestic and international actors, resource depletion, population growth and climate change. These may appear to be issues which are far more relevant to the Global South, and less relevant to the Global North where governments are expected to provide a framework for us to meet our needs through selling our labour on the market and buying what we need on the market. However, we are increasingly becoming aware that such a system not only has huge environmental costs, but is also highly divisive and economically unstable.

It is in this context of larger threats – in particular climate change, resource depletion and the potential for financial meltdown – that a third force has emerged which is increasingly pushing for the restoration of commons regimes in the Global North and South. It often begins – as the Transition Town movement began – by trying to reduce a communities' impact on the environment through trying to build its local resilience (its capacity to meet its needs itself), but as such initiatives seriously try to grapple with re-localisation they inevitably come up against those forces which appropriated the commons in the first place.

Such movements see humans as being most at home in 'commons' rather than in 'capitalist' or 'command and control' social forms. They assume that we will be far wealthier with less if we end the extraordinary levels of inequality and if we ensure that we all have sufficient to meet our needs, at the same time ensuring that we do not destroy the environment on which we and our children's futures, *let alone* the future of other species, depends.

3. Can we apply commons principles to the global climate commons?

While use of community commons is regulated by members of the community to ensure that resources are available and replenished for current and future generations, there is clearly a severe problem with the fact that the atmosphere and skies are a commons that are not local, that are essential to all, and that are being polluted and destroyed by some actors to the detriment of all.

The current global response to this predicament focuses on establishing international institutions that can develop, oversee and bring sanctions to bear on countries and corporations who are responsible for – for example – emitting pollution that is causing temperatures to rise year on year, something which the science recognises will become an unstoppable process if not rapidly dealt with. The difficulty is that these international institutions – for example the UNFCCC – are dominated by governments and corporations who have a (short term) vested interest in an ever expanding economy that increases rather than decreases the throughput of carbon from fossil fuels through the economy and into the atmosphere.

Chapter 5 examines whether and how a 'Global Climate Trust' based on a 'Climate Charter' could provide the organisational means for us to be able to protect the Global commons, and similarly Peter Barnes has proposed a 'Sky Trust' to deal with this issue.

This chapter seeks to outline how Cap and Share could be part of a commons approach to the climate and related crises, and how it is likely to be introduced not through government diktat or international agreement but through individuals, communities and networks pressurising their individual governments to adopt such an approach – perhaps in order to address the climate crisis but more probably as part of a strategy designed to rapidly reign in finance as people insist on a different response to the extortions of the financiers.

> Elina Ostrom writes that:
>
> "Currently, efforts to address climate change are being orchestrated primarily by global actors, but waiting for international solutions is wasting valuable time. Conventional wisdom tells us that there are only two options to deal with managing resources: either privatization or management by the state. This view is hindering progress. To successfully address climate change in the long run, the day-to-day activities of individuals, families, firms, communities, and governments at multiple levels—particularly those in the more developed world—will need to change substantially. Encouraging simultaneous actions at multiple scales is an important strategy to address this problem."[16]

Although Ostrom's writing is very helpful, it is perhaps a mistake to think that "efforts to address climate change are being orchestrated primarily by global actors". Global actors are exactly those who are failing to address climate change, and very often deliberately so, since addressing it would mean reigning in the economic system driving climate change, the very system that has given 'global actors' their prominent position. Rather, efforts to address climate change are happening at the very local level and in peoples' movements and networks that are also trying to reign in the system driving climate change.

Part five of this chapter will examine the proposition that Cap and Share is most likely to be introduced in particular places as part of a package in which communities are seeking to defend or restore commons ways of limiting resource use so that all can have sufficient to meet their needs. Indeed, this is how radical change has always happened: it happens through example, through the leadership of innovative people not through the imposition of government or international agreement.

Having a global framework in place (for example the 'Global Climate Trust' based on a 'Climate Charter' as proposed in Chapter 5) is crucial in order to provide an essentially non-governmental global framework to enable this radical grassroots and country-by-country reclaiming of the commons to converge at the global level. The network of people with ecological values that Jopling writes about in Chapter 5 already exists, what is needed is for such networks to strengthen in depth and breadth. Jopling writes that:

> "This [Global Climate Trust] initiative can only take off if there are people sharing the same values around the world to make it happen. We are confident that there are. The participants in this task will be the millions of people around the world who understand the peril we are in and who put their responsibilities as world citizens ahead of other loyalties and interests."

Perhaps, though, such people will act to do this, less out of a desire to "put their responsibilities as world citizens ahead of other loyalties and interests" but more out of a desire to ensure their localities are resilient and able to meet their and their children's needs. For this resilience to be defended, restored or maintained requires change at a society-wide and global level – but the impetus for this social and global change may well come from the very immediate need to feed families and ensure the safety and well-being of communities in their localities.

Part five of this chapter will return to the movements for land reform and re-localisation in the Global North, and examine the role that Cap and Share could play in their attempts to reclaim their localities and thereby (wittingly or unwittingly) defuse the causes of climate change and resource depletion. The next part, though, focuses on the Global South.

4. Redirecting policy to defend Commons Regimes from Climate Change 'Solutions'

This section argues that the most efficient and reliable way to ensure that carbon emissions from deforestation in the Global South are stopped is not to pour a vast amount of public money into establishing policies like REDD (Reducing Emissions from Deforestation and Degradation). It is quickly becoming apparent that REDD will be dependent on a carbon trading scheme that is precarious anyway because carbon trading is levelling off. Worse, it is likely to systematically exclude the very people who have been protecting their forests for generations. Instead, it would be better to use public money to directly support the recognition of customary land tenure and the protection of sustainable land use in the forests of the Global South by local people.

Across the Global South indigenous peoples commons systems are being threatened by enclosure, and potentially by policies such as REDD (Reducing Emissions from Deforestation and Degradation).

Given the refusal of governments and corporation to make the necessary dramatic cuts to emissions in industrialised countries and sectors, the dominant policy response to Climate Change since Copenhagen has been to focus on REDD.

Tackling the causes of deforestation would be a welcome move both globally and from the point of view of those peoples – most of them in commons regimes – who rely on the forests for their livelihoods. However REDD, does not address the real drivers of deforestation. Instead it would enable polluting corporations in the North to buy carbon credits from governments and institutions in the Global South which can 'prove' they are reducing emissions from forests and thereby gain the right to sell carbon credits. The credit is calculated from the difference between projected deforestation rates and the actual rates that are supposed to follow when these governments and institutions intervene to protect forests. However these interventions are based on the totally mistaken notion that it is local people who are the primary drivers of deforestation. So proving that governments and institutions are stopping them using their forest, means that they can claim to have reduced deforestation and thereby gain carbon credits to sell.

It has long since been proven that the drivers of deforestation are not poor farmers, pastoralists or hunters and gatherers but the clearing for large scale logging, commercial farming, ranching and plantations which force local people off their own land[17]. Often, once forced off their land by these concessions and protected areas, they are unable to use the forest resources sustainably as they have in the past (which is why the forest was still there). Unable to pursue small scale rotational shifting cultivation, or small scale harvesting from the forest, they can then indeed become part of the process that is destroying the forest, whether through working for the concession holder or extracting what they can from the resource before these more powerful others destroy it completely. However it is important to differentiate between this reaction in pursuit of some form of livelihood for their families and the driver of deforestation that needs to be checked. Securing forest peoples land tenure and requiring real accountability, transparency and the necessity of companies negotiating with local people on an equal playing field would begin to address this.

In contrast, the McKinsey Report's cost curve has famously sought to compare how much it costs to stop deforestation in different contexts in a way that suggests that stopping local people costs little but has a high benefit (so this is the option that should be pursued) whereas the cost of trying to stop a palm oil plantation, for example, is very high but the benefit is not huge[18]. This is a clear example of preferring to address a symptom in a way which fuels the cause: further marginalising local peoples' ability to use their forests sustainably, and further strengthening the hand of those who cut the forest to create permanent industrial plantations.

In any case – the weakness of the carbon market means it is no solution [19]

An additional problem for REDD, based on the carbon market, is that the latest World Bank Carbon Market Report paints a clear picture of a global carbon market that is levelling off, and has failed to reduce emissions. The global carbon market is worth $145 billion, but 97% of this is either the EUTS (the European Union Trading Scheme) or led by EUTS credits. In effect, there only really is the EUTS carbon market, and it explicitly excludes forests because to include them would make far too many credits available in its system. REDD isn't in there, and can't be until at least 2020 when the second period of the EUTS finishes.

A second point is that the voluntary carbon market is worth a tiny fraction of this amount, $450 million, of which only 30-40% relates to REDD (roughly $145 million) and half of the people investing in the voluntary market have only been doing so because they are expecting the compliance market will take off. If they wish to pursue the chimera of a global carbon market, the UK and others are going to have to continue pumping money into creating the framework for a market that is levelling rather than taking off.

To understand why the market is failing it is useful to look briefly at the European Union Trading Scheme on which it (almost) entirely depends.

The EUTS is failing in terms of actually reducing carbon emissions, partly because €2 billion worth of permits were given for free to the largest and most polluting companies while the renewable companies received nothing. Looking back, carbon traders say these permits should have been auctioned. The reason they weren't was entirely due to the fact that the most heavily polluting companies had too much clout. This has had the consequence that other measures – which could have been undertaken to actually reduce emissions – are unable to be introduced because they are seen as potentially interfering with the EUTS.

From a carbon trading perspective, the EUTS simply doesn't work. It has given out far too many permits, and has proved itself very susceptible to many different types of frauds – from hacking into member states' systems, to the stealing of permits, to people selling permits twice. Thus there is no delivery of reductions, no security of price, and no security in the system. The EUTS has been described as being successful only at: dramatically increasing the profits of the largest polluters; increasing energy prices for consumers; and increasing emissions.

The broader political context is shifting rapidly

People in the global north have watched as their Governments have poured their public money into the major financial institutions. These institutions were on the brink of collapse because they were trading in bundles of debt where no one could distinguish between debt that might be repayable and debt that could never be repaid.

To pour public money into setting up the infrastructure to enable financial institutions to trade in a commodity, carbon permits and credits, the very existence of which entirely depends on being able to make a convincing distinction between schemes in terms of the degree to which forest loss is projected to be less than expected due to that schemes' activities, would seem very unwise to say the least.

In terms of deforestation and tackling emissions there are three clear reasons why this is unwise.

Firstly, the money REDD schemes make arises from the difference between projected deforestation and the rate of actually occurring deforestation. One of the easiest ways to intervene to try to prove they are lowering the rate of carbon emissions projected for a particular forest is by visibly stopping local people making use of their customary resources.

Secondly, this is an approach that demonstrably drives deforestation rather than curbs it. In Guyana the baseline projection of business as usual assumed that 0.45% of the forest would be deforested each year, (even though when the estimate was made only 0.02% was being deforested each year). This meant that Guyana could increase deforestation 20 fold and still get REDD money[20]. Clearly with REDD payments, the worse you can portray yourself as being the more payments you can get, and if you have been and intend to continue being a good custodian of your forest then you will get nothing. Dyer and Counsell comment that "there is a danger that these high-deforestation scenarios could become self-fulfilling prophecies and encourage tropical governments to pursue destructive practices in order to increase their expected compensation"[21]. The incentives in the system have perverse effects.

Thirdly, just as the EUTS distracts us from far more effective measures to reduce emissions in Europe, REDD distracts us from the need to address demand reduction in the Global North. Instead the prospect of REDD, the existence of the EUTS and a focus on building the carbon market in general, enables countries and industry in the Global North to avoid having to make the dramatic cuts required by buying permits/ credits. The fact that these permits/ credits seek to 'prove' that a REDD scheme is reducing the level by which global emissions are INCREASING means that this whole approach is logically not about making deep cuts but about reducing the speed at which global emissions are rising, while providing a means by which countries and corporations in the Global North can postpone making the deep cuts themselves.

What concerns us here is not just the logical impossibility of a carbon market system achieving it's stated goal of helping to decrease global carbon emissions[22]. That is increasingly obvious to even the most central institutional players and is something that should be thrashed out between citizens and their governments in the global north. What concerns us here is also the impact of this system on people in the global south.

What forest people in Cameroon ask is: why don't you – instead of pushing REDD – address the cause of these emissions: your industries in the north?[23]

They would like public money from the global north to help them to ensure that their ownership of their forests is recognised – in other words to secure their customary land tenure systems – so that they can continue to manage, protect and sustainably use their forests. In the enduring absence of such offers from the global north, many will also be willing to gain some forms of development compensation in return for losing their forests to conservation and other actors, since otherwise they think they will be left with nothing: neither rights to their land, nor compensation for those rights being denied[24].

The amount of public money from the global north that is required to support communities in the global south to do this (to achieve security of tenure over their customary forest lands, and to protect or establish sustainable ways of their benefiting from being in the best position to protect those forests), the amount of money required for this is finite, laws once changed can endure, and the projects required are visible, tangible and of immediate benefit in helping to end poverty and support democratisation in some of the poorest countries of the world.

The evidence is amassing (from World Bank[25], academic[26] and NGO studies) that the people best placed to protect forests are forest peoples' themselves.

Clearly the cheapest, most efficient and most reliable way to ensure that happens is not to pour a vast amount of public money into establishing a highly questionable commodity and a precarious trading scheme based on projects that are likely to systematically exclude the very people who have been protecting this forests for generations. The best way to use that money is instead to directly support the recognition of customary land tenure and the protection of sustainable land use in the forests of the Global South.

Solution: Redirecting current finance to support commons regimes

The situation is ripe for a re-examination of how public finance is to be directed. Such a re-examination may discover a much worse situation in terms of the consequences of current policies, but a much better situation in terms of the consequences of refocusing finance on direct rights based action to supports modern customary rights and highly adaptable sustainable local use of resources to alleviate poverty in the present.

This does not require an international framework and agreement to get going. It can start from single countries – or indeed cities, regions or towns – in the Global North taking this approach and building a coalition for action in the present. Rather than having to wait for international agreement on a particular financial mechanism that all have agreed on, public finance can happen now and – seen as part of overseas aid budgets – can come from any public source without having to await a seemingly unreachable Global agreement.

From a UK perspective, this is simply about carefully targeting the £2.9 billion the UK has committed to the International Climate Fund for climate related work outside the UK over the next 4 years.

Governments in the Global North are likely to increasingly find it hard to justify spending public money on seeking to reduce global emissions by pumping more and more money into trying to create a global carbon market. In this context, where co-ordinated and effective international action on climate change is unlikely to materialise, what is clearly needed is bold, cost effective leadership from the UK that can involve directing the money we have already committed to this issue in a way that can achieve transformative change.

In the process the UK can help build sustainable systems under local control that can stop deforestation, raise much needed revenue for their governments, and directly address the need to secure carbon not by bringing in companies to clumsily interfere, and incapably count carbon, but by ensuring forest peoples rights to their commons are recognised and they are able to preserve their forests as the basis for their development choices.

5. Restoring the commons, and how Cap and Share is a crucial part of this process

Action in the Global South does not need to wait for a global agreement but can be embarked on by governments, regions, cities or communities in the Global North making common cause to support communities in the Global

South to defend their commons. Similarly action in the Global North can be enabled unilaterally at a society-wide level, partly through people pushing for their governments to adopt policies such as Cap and Share.

We need to rapidly de-couple our everyday activities from the economic-industrial complex by enacting an alternative – re-creating commons regimes here like the ones that currently provide sustenance and social organisation for between 1 and 2 billion people elsewhere in the world.

Whether *proactively* in the context of a booming capitalist economy, or whether *reactively* in the context of the IMF threatening to pull the plug on loans to Scotland or the UK; whether in order to stave off impending ecological collapse and climate change, or whether to resist blackmail and extortion when the IMF approaches us in the same brutal form it approaches other parts of the world: the underlying paradigm of the commons can enable us to recognise that we are *not* that ecocidal machine, that we *can* self-organise, and it can remind us how.

To enact a powerful alternative in the Global North, we need to prefigure the commons at a community level, we need to engage in civil disobedience to stand up to those forces that claim there is no alternative to business as usual, and we need a set of proposals, a route-map, to help us disentangle ourselves from the ecocidal machine we have been convinced we are wedded to. Such a route-map needs to work both within the electoral system as it stands (within a 'successful' expanding capitalism, within a party political voting system, within the economic inequalities that exist) and also work within a rapidly transforming situation where people will be rapidly losing faith in the status quo.

At the community level: We need to work to expand the land reform movement from rural to urban areas and we need to work in Transition and other positive local initiatives to create resilient need-meeting structures.

Transition Initiatives as Lifeboats:

There are a huge range of re-localising initiatives in Scotland: from Transition Towns, to Land Reform initiatives or Development Trust Associations – all focused on regaining local control of local resources. Transition initiatives, for example, often seem to exist in a parallel world – developing local currencies, mapping food growing areas, developing energy sources – sketching a different re-localised way of meeting our needs as a result of a strong sense that this system cannot continue.

When you are on a ship that is ploughing full steam ahead, the existence of lifeboats perched awkwardly above the side decks seem completely irrelevant to everything you are doing. If the ship starts to sink then suddenly they are all that matters.

Transition and other similar relocalisation initiatives are like lifeboats being developed on the side of this economic oil tanker. Huge effort is going into keeping economic growth going full steam ahead, other less powerful efforts are going into trying to turn the tanker round, to develop a steady state economics that stops depleting the earths resources. Still others say that we cannot turn it round fast enough; that it is heading straight for the rocks of ecological meltdown. And then suddenly the tanker starts developing gaping holes above and below the water line. Desperate attempts are made to patch it up, billions are poured into a failed financial system: billions and billions are poured through the holes in the ship. What if the ship is holed irreparably? If the lifeboats are ready then the ship being holed may be a lifesaver for us all. Whether there is enough food and water and strength for us to help each other reach the shore, the direction of travel is now entirely different.

Engaging in civil disobedience: Here in Edinburgh, as in hundreds of cities across the world, people have for weeks now occupied a public space and are camping out in it, handing out leaflets, holding discussions and showing films about how the financial system is driving savage social inequality and ecological destruction. The traffic thunders by, and passers by walk on or pause at this collection of tents in St Andrews square where there is as much focus on process – on the challenges and possibilities of living together – as on campaigning. One man who described having slept on other peoples couches for years, described his experience like this:

> "I feel at home. We're having an effect. We're doing something. It's the end of bitterness for me. Waking at 4 in the morning on a Friday night to hear the drunks climbing over the fence and being met by the love police [members of the group dealing with security, with 'love police' emblazoned across the back of their fluorescent jackets], and just hearing them talk about the state of the world, it makes it worthwhile".

Like the UK Uncut movement, the Occupy movement has mobilised huge public sympathy behind their insistence that the 1% – the financiers who gamble with our wealth and hoard it for themselves – should also be paying their way, rather than pursuing their addiction to a system that extracts wealth and destroys the social and natural world. Following on from the mass mobilisations in Tunisia and Egypt, from the huge gatherings in Syntagma Square in Athens, from hundreds of thousands taking to the squares in entirely unexpected, creative non-violent protest in Spain, from Occupy Wall Street, perhaps we need to be prepared for the fact that as the ecological crisis manifests itself through economic meltdown, there is the opportunity for an extraordinary reclamation of democracy, including through ways of organising protest that demonstrate and seek to prefigure and bring into being the kinder society we want and need.

In the Occupy movement there has been a healthy refusal to be narrowed down to a set of demands, an insistence that we all need to face up to what has happened and find a way through, and in Oakland and elsewhere there are moves towards occupying other spaces where we can care for each other and engage in productive work. That is not to say that there is no need for a clear, albeit flexible, routemap to chart our way out of this mess, it is just to say that the vastness of the transformation required cannot be reduced to a single solution.

For example, at the society-wide level in Scotland: we to need to communicate a clear set of proposals that can reshape the economic and political system so that it supports, rather than blocks, the resilience building happening at the community level, and so that it creates the basis for radical change. There are very clear policy steps that can be taken, including:

i) Refusing to recognise agreements reached in tax havens, mutualising and regulating **financial** services, and ensuring that social and environmental good rather than shareholder dividend is the legal bottom line;

ii) Supporting the development of **community energy** and community food initiatives;

iii) Deciding **planning** applications not on the basis of economic growth but on the basis of whether a development contributes to community resilience;

iv) Ensuring that deepening **self-determination** for Scotland becomes self-determination for communities, including through fundamental land reform that returns power to communities; and

v) Unilaterally developing **Cap and Share** for Scotland.

These last two are the crucial game changers.

In terms of self determination: Being for independence need not be about drawing a boundary on a map and ignoring the rest of the world, it can be about asserting the right of people in a self-defined area to determine their own affairs in a way that enhances the independence, self determination and autonomy of others. People on the Isle of Eigg worked to ensure it gained its 'independence' and inspired others to do likewise. People in the Transition initiative in Portobello, Edinburgh, are working to develop local resilience (to halt our destruction of the ecosystem, and to prepare for huge economic collapse) and have helped inspire other communities to do likewise. Scotland can play a similar role in the world: when small places trail-blaze, others can recognise a route they want to follow.

Cap and Share is radical because it takes the climate problem more seriously than governments and corporations do, and is realistic because it is a new way of applying an old, enduring approach to self organising society: the commons. It recognises that we all rely on a stable climate and must treat it with care. It reflects two key aspects of commons regimes:

> **Firstly**, the shared managing of resources in a way that enables all right-holders (all members of a community) to benefit from the resource, and to limit their use of the resource so that all others may continue to benefit from it.
>
> **Secondly**, although such commons agreements may now need to be embodied in legislation to constrain more powerful actors, their origin, power and legitimacy arise from people, communities and countries unilaterally adopting this approach and recognising that this approach helps to resolve a myriad of other interrelated issues that threaten our well-being.

Most human societies have self-organised the allocation of resources in this way[27], and in our society such a commons approach to managing resources continues within the family, amongst friends and neighbours, in mutual societies, and in all those aspects of life where the calculus of self-interested money-making has not been imposed – as if it were natural – by the ideology of the Market backed up by the force of the State.

This ongoing imposition of that Market ideology is evident in relation to areas often previously considered inalienable common goods – in the UK the health service and public forests, in the Global South privatised forests, genes and water.

However, in the last year large parts of global society have gone into revolt in a whole swathe of public demonstrations that have sought – and often succeeded – in bringing down governments, whether in Tunisia or Egypt, or in Athens and Madrid. These revolts are examples of an ongoing gathering storm that has swept out neoliberal governments across Latin America over the last decade, and seen the rise of a powerful indigenous movement across the world.

Both movements have begun to intersect not just in places like Bolivia where an indigenous government has rejected the useless market response to the climate crisis but also in a Global movement to protect and reclaim the commons – whether evident in the Canadian Cree protecting their forests from loggers, Indian farmers stopping their land being turned into car factories, or Scottish crofting communities regaining collective ownership of their common resources including land.

Up against this movement is the ideology that insists that growing the economic cake is the only possible way of maintaining public order as the number of people who share it increases. Yet if economic production is divided fairly its size can be reduced. In other words: if the amount of extraction and consumption, of industrial activity and resulting emissions, can be reduced the vast majority can still experience their living standards as increasing. They can balance their needs with others and with the environment in a way that ensures the greatest possible sense of well being: the assurance that our children's lives will not be ravaged by the consequences of climate change, and the assurance that our actions in the present are not ravaging other humans, other species, and our shared ecosystem.

In 'The Spirit Level' Richard Wilkinson and Kate Pickett[28] provide the evidence to demonstrate how a more equal society benefits all and a more unequal society impoverishes even the wealthiest. We should extend their argument to include future generations and people in other societies.

This is where Cap and Share fits into the bigger picture – it can establish itself within this electoral and market system yet help us bridge to a radically different world: that of the commons.

In Scotland we have been proposing that Cap and Share works in its "Cap and Dividend" version as follows: those bringing carbon into the economy (those very few companies importing or producing coal/ oil/ gas/ cement, etc.) would take part in an annual auction to buy the right to bring carbon into the economy. **The extra price** they have paid would then be **passed on to manufacturers** and other users of the fossil fuels. This leads to higher prices for all those using products, services, modes of transportation, etc., which have carbon embedded in them.

This means that those using more than their fair share of carbon are penalised because all such prices will have risen, while those using less (the great majority) will benefit either with extra cash or with other ways of meeting their needs (see below). The rapid rise in the price of fossil fuels will mean that producers will be encouraged to rapidly develop non-carbon based products/ services/ modes of transport, and avoid producing carbon ones.

Meanwhile, **the cash generated** from the auction of these carbon emission permits would be **passed on to the population at large**:

- In one version it would be passed directly into peoples' bank or post office accounts so that they can deal with the increase in prices.
- In another approach such cash may well be passed on to individuals but could also – to a greater or lesser extent – be used to enable society wide shifts to energy healthy infrastructure, and to support the development of community resilience.
- In yet another version, each individual receiving their right to the money generated by the acutioning of permits to import carbon, could – instead of taking the money – decide to forgo the money and so stop that portion of carbon from being allowed to enter the economy. In this way a small minority of farsighted people could unilaterally help

rapidly reduce the amount of carbon passing through the economy. An aproach which powfully echoes the way that commons systems enable individuals to have a disproprtionately beneficial impact when they are acting in the long-term interest of all, while contraining the destructive impact of those who do not share that concern.

In terms of reducing carbon emissions, this policy would ensure that high-carbon products, modes of transport, energy sources and services, are fast replaced by zero-carbon ones. The necessary rapid rise in the cost of high-carbon options would be accompanied by the rapid development and shift to zero-carbon ones, thereby dramatically reducing and then *stopping carbon being extracted from the ground to pass through the economy into the atmosphere*.

In a market economy that is functioning reasonably well, the version to push for might well be the first one – in which the tariff on fossil fuels coming into the economy is redistributed directly to peoples bank accounts, leaving the vast majority of the population immediately better off and only those who can afford it (the heavy emitters) penalised for disproportionately polluting the global commons. In this context, this 'Cap and Dividend' version would be electorally attractive, and would be immediately workable since it would start by regulating rather than replacing the powerful companies responsible for bringing carbon into the economy.

In a market economy that has seized up, where financiers are unable to cash in on the ponzi schemes they have been selling, the second version might well be far more appealing. In this version, much less of the tariff on fossil fuels coming into the economy is redistributed directly to peoples bank accounts, and much more is used to support communities to build local resilient need-meeting structures. In this context, food and warmth and care will be seen as far more reliable and necessary than a fluctuating and untrusted currency.

Key advantages of cap and share:

Cap and Share doesn't create a different political or economic game, it simply changes the rules of the game so that the market has to internalise the carbon cost. However, in doing this, it can act as a bridge towards a very different paradigm: that of the commons where the people are recognised as sharing the atmospheric resource and responsibility for it. Meanwhile, the certainty of the cap (the permitted level of carbon in the economy) being reduced rapidly year on year, would immediately boost jobs and investment in the development of zero-carbon energy, goods and services. Together with the Government reducing energy demand and ensuring 100% renewable energy, and re-regulating and re-directing finance, Cap and Share helps create a level playing field in which food and energy, goods and services, will be produced closer to home, helping build socially and ecologically healthy communities

Key challenges of cap and share:

UK law would not allow Scotland to unilaterally implement such a system, but if people in Scotland powerfully push to do so, this would put huge pressure on the UK Government to follow suit or to accept Scottish autonomy in this and related areas. Similarly, the EU and WTO could argue against us imposing stringent tariffs on carbon embedded imports from countries that do not have a similar scheme. However, such a tariff would be necessary to create a level playing field by levelling up international practice. Insisting on this would need to be connected with an international movement, but unilaterally pursuing this policy would be one way of making such a movement's presence felt, and one way of disentangling ourselves from the Ecocidal paradigm and practices of the current Market.

Impact of this unilateral re-localisation process on the Global South:

Cheap products from the Global South would be priced out of our market through the requirement that they internalise their carbon costs or, if not, have tariffs imposed on entry. However, although re-localising here would hit the ability of the elites' in the Global South to sell cheap produce to the Global North, it would also hit their ability to take peoples' land in order to grow cash crops to sell to the Global North. This applies as much to the power of the elite in China to control manufacturing, as it does to the elites in Africa, Asia and elsewhere who expel communities from their land in order to produce the raw materials – timber, palm oil, minerals, etc – which such globalised manufacturing depends on[29]. Since peoples wealth is ultimately grounded in community control of local resources, and if people only become dependent on those who control the Market when their land is taken and they are turned into landless labourers, then our insistence on re-localising here, including through implementing Cap and Share, becomes a statement of solidarity with the world's poor. What they want is for us to take the power away from the corporations that back their elites. For them, as for us, protecting and restoring commons based systems that meet peoples real needs, and defending these against the land-grabbing forces of enclosure, is the same struggle whether in the Global North or South.

When activists from the Global North went to the Philippines to support local peoples struggle against the building of a dam that would have flooded their land, the activists were not asked "What can you do for us?" but "What is your struggle?"

Cap and Share can enable societies of the Global North to reduce our dependence on fossil fuels, and to enable us to move on to sustainability through re-localisation grounded in radical land reform processes. Communities in the Global South need us to be using policies like Cap and Share to tackle at source the root causes of a process that is dispossessing us all; and more immediately they need support in protecting their commons, they need defence against those who threaten to dispossess them, and solidarity in demanding the return of their lands from those corporations and powerful elites who have already dispossessed them.

The same machine, malaise and misunderstanding drives both the ecological and economic crises. We need to respond by defending and restoring commons systems of mutual care.

What looks radical now will be overtaken by events. Can we be ahead of those events and offer action, policies and a story that speaks to who we really are and what we really need?

Endnotes

1. Anderson, Kevin and Alice Bows 2008 'Reframing the climate change challenge in light of post-2000 emission trends', *Philosophical Transactions of the Royal Society* A http://rsta.royalsocietypublishing.org/content/366/1882/3863.short
2. Anderson, Kevin 2009 'Reframing climate change: 'impossible' challenges. Mitigate for 2C, while adapting to 4C'. Paper given at the University of Glasgow
3. Point Carbon News 2011. EUAs slide towards 9 euros, hit fresh 33-month low 18 Nov 2011, in Point Carbon News http://www.pointcarbon.com/news/1.1683984 (accessed 23rd December 2011)
4. J B Priestley 1940. Postscript Sunday 21st July 1940, in the Special Collections of the University of Bradford. http://specialcollectionsbradford.wordpress.com/2010/07/21/postscript-sunday-21-july-1940/ (accessed 23rd December 2011)
5. Alden, Wily, L (2011) *Accelerate legal recognition of commons as group-owned private property to limit involuntary land loss by the poor*, January 2011 International Land Coalition, Rome. www.landcoalition.org/publications/accelerate-legal-recognition-commons-group-owned-private-property-limit-involuntary-lan
6. Alden Wily, L (2011) *'The Law is to Blame' Taking a Hard Look at the Vulnerable Status of Customary Land Rights in Africa* Development and Change Volume 42 (3) May 2011.
7. Alden, Wily, L (2011) *Whose Land is It? The Status of Customary Land Tenure in Cameroon*. CED, FERN & RFF, Yaounde & London. www.fern.org/sites/fern.org/files/cameroon_eng_internet.pdf
8. Jay Walljasper (2009) Tragedy of the Commons R.I.P. On The Commons (13th October 2009) http://www.stwr.org/economic-sharing-alternatives/elinor-ostroms-nobel-prize-tragedy-of-the-commons-rip.html
9. Ostrom, Elinor *et al.* (2002) *The Drama of the Commons*. National Academy Press, Washington DC
10. Kirkby, John, Phil O'Keefe & Lloyd Timberlake eds. 1995 'The Commons: where the community has authority' in *The Earthscan reader in sustainable development*. London: Earthscan.
11. Walljasper, Jay (2011) The Victory of the Commons may 16th 2011 http://activism101.ning.com/profiles/blogs/the-victory-of-the-commons
12. Kenrick, Justin (2006) 'Equalising Processes, Processes of Discrimination and the Forest People of Central Africa', in T Widlock & W Tadesse (eds.) *Property and Equality: Vol. 2 Encapsulation, Commercialization, Discrimination*. Oxford: Berghahn
13. Ostrom, Elinor (1990). *Governing the Commons: The Evolution of Institutions for Collective Action*. Cambridge University Press. ISBN 0-521-40599-8.

14. Wightman, Andy (2010) The Poor Had No Lawyers. Who owns Scotland (and how they got it) Edinburgh: Berlinn
15. Alden, Wily, L (2011) *Whose Land is It? The Status of Customary Land Tenure in Cameroon.* CED, FERN & RFF, Yaounde & London. www.fern.org/sites/fern.org/files/cameroon_eng_internet.pdf
16. Ostrom, Elinor 2010. A Multi-Scale Approach to Coping with Climate Change and Other Collective Action Problems. Solutions. Vol 1, No. 2. pp. 27-36 – http://www.thesolutionsjournal.com/node/565
17. Boucher, Doug *et al* (2011) The Root of the Problem: What's driving tropical deforestation today? Union of Concerned Scientists http://www.ucsusa.org/global_warming/solutions/forest_solutions/drivers-of-deforestation.html
18. Greenpeace (2011) Bad Influence: How McKinsey inspired plans lead to rainforest destruction http://www.greenpeace.org.uk/media/reports/bad-influence-how-mckinsey-inspired-plans-lead-rainforest-destruction
19. Kill, Juta *et al* (2010) Trading carbon: How it works and why it is controversial. Fern: London. http://www.fern.org/tradingcarbon
20. Global Witness (2011) Guyana sees a three-fold increase in deforestation despite landmark deal to protect its forests (7th March 2011) http://www.globalwitness.org/library/guyana-sees-three-fold-increase-deforestation-despite-landmark-deal-protect-its-forests
21. page 6 Dyer, Nathaniel and Simon Counsell (2010) McREDD: How McKinsey 'cost-curves' are distorting REDD. Rainforest Foundation UK – Climate and Forests Policy Brief November 2010 http://www.rainforestfoundationuk.org/McRedd_News
22. Munden, Lou (2011) REDD and Forest Carbon: Maket based critique and recommendations. http://www.mundenproject.com/; see also: http://www.redd-monitor.org/2011/03/22/munden-project-report-on-redd-and-forest-carbon-forest-carbon-trading-is-unworkable-as-currently-constructed/
23. Forest Peoples Programme (2010) Cameroon REDD Community Consultations and Civil Society Workshop. http://www.forestpeoples.org/topics/redd-and-related-initiatives/news/2010/07/press-release-cameroon-redd-community-consultations
24. Forest Peoples Programme (2011) The indigenous peoples of Cameroon: from Ngoyla-Mintom to national recognition http://www.forestpeoples.org/topics/redd-and-related-initiatives/news/2011/10/indigenous-peoples-cameroon-ngoyla-mintom-national
25. Nelson, Andrew and Kenneth M. Chomitz (2011) "Effectiveness of Strict vs. Multiple Use Protected Areas in Reducing Tropical Forest Fires: A Global Analysis Using Matching Methods," *PLoS ONE* 6, no. 8 (2011): e22722. http://www.plosone.org/article/info%3Adoi%2F10.1371%2Fjournal.pone.0022722
26. Porter-Bolland, Luciana *et al.* (2011) "Community managed forests and forest protected areas: An assessment of their conservation effectiveness across the tropics," *Forest Ecology and Management* (June 2011), http://www.cifor.org/nc/online-library/browse/view-publication/publication/3461.html
27. Graeber, David (2001) *Toward an Anthropological Theory of Value: The False Coin of Our Own Dreams.* New York: Palgrave
28. Wilkinson, Richard and Kate Pickett (2010) The Spirit Level: Why More Equal Societies Almost Always Do Better Alen Lane: London
29. Oxfam (2011) Land and Power: The growing scandal surrounding the new wave of investments in land. 151 Oxfam Briefing Paper 22 September 2011

Chapter 3

Cap and Share in Pictures

Laurence Matthews

This chapter illustrates some features of Cap & Share (C&S) and its variants. Pictures (each worth a thousand words, after all) can be useful for understanding C&S and for explaining it to others.

Cap & Share (C&S) is a way of limiting the carbon dioxide (CO_2) emissions from burning fossil fuels [1]. It could operate on a global scale, but to start with, imagine a national scheme applied to a single country's economy.

As the name implies, there are two parts to C&S:

Cap: The total carbon emissions are limited (capped) in a simple, no-nonsense way.

Share: The benefits, from the huge amounts of money involved, are shared equally.

These principles are simple, but there are a number of ways of interpreting them, especially the 'Share'. The principles are more important than the details, so let's look first at the main principles themselves.

The Cap

The cap is a limit each year on overall, total CO_2 emissions. This is set, in line with scientific advice, at a level which will bring concentrations of CO_2 in the atmosphere down to a safe level. But how do we ensure this cap is met?

The 'Cap' in Cap & Share is 'upstream'. A good way to explain this [2] is to think of watering a lawn with a hosepipe connected to a lawn sprinkler, with lots of small holes spraying water everywhere. If you wanted to save water, what would you do? One idea is to try to block up all the sprinkler holes one by one. But wouldn't it be simpler to turn off the tap a bit?

It's the same with fossil fuels, where the sprinkler holes correspond to the millions of houses, factories and vehicles 'downstream' (on the right-hand side of Figure 1), each emitting CO_2 by burning these fuels. But corresponding to the tap (on the left-hand side of Figure 1) there are only a few primary fossil fuel suppliers (e.g. oil companies) who introduce fossil fuels into the economy – by importing them or extracting them from the ground.

Figure 1

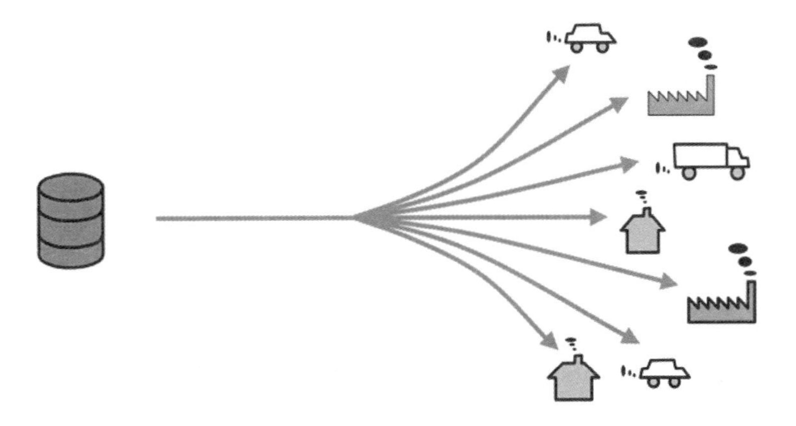

So it's easier to work 'upstream', on the left-hand side of Figure 1. The cap operates by requiring the fossil fuel suppliers to acquire permits. A permit for 1 tonne of CO_2 entitles a fossil fuel supplier to introduce fossil fuel with that CO_2 content – that is, the amount which will emit 1 tonne of CO_2 when burnt. The number of permits issued equates to the desired cap. By controlling the supply of *fossil fuels* coming into the economy, we automatically control the *emissions* which occur when those fossil fuels are burnt somewhere down the line.

For imports of fossil fuels, the cap may best be applied at the point of entry to the country (port, pipeline); for domestic production the best place is likely to be the mine (for coal) or the refinery (for gas/oil). The details need to be sorted out, but will be able to make use of systems already in place for accounting and taxation purposes, and will be in any event much simpler than trying to account for everything downstream.

Now look at Figure 2 overleaf. This shows the two halves of our carbon footprint ('direct emissions' and 'indirect emissions'). Some fossil fuels are burnt by companies producing goods and services on our behalf – these are our 'indirect emissions' and form the upper line in Figure 2. The goods that reach us have these emissions 'embedded' in them. Meanwhile we also buy fossil fuels (such as petrol) directly and burn them ourselves – these are our 'direct emissions' and form the lower line in Figure 2. The permits are represented by the small rectangles. Direct and indirect emissions are both taken care of by the same system.

Figure 2

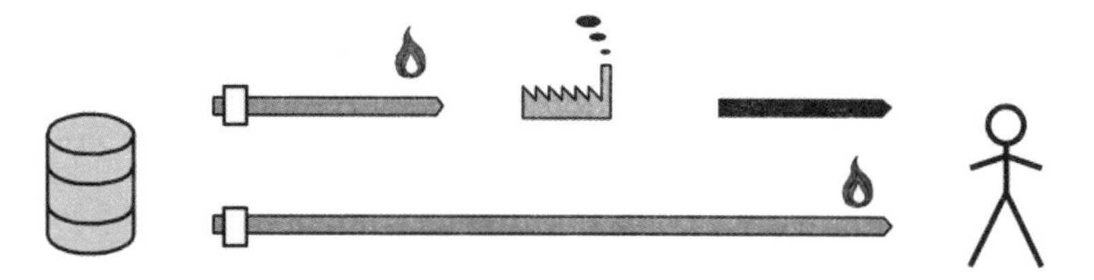

In other words, it doesn't matter where the emissions take place (the emissions are indicated by flame symbols in Figure 2). We don't need to monitor and measure emissions at all, because instead of focussing on the emissions, we are focussing on the fossil fuels themselves. This focus also makes clear the connection between controlling emissions and 'keeping fossil fuels in the ground'.

(Elaborations to this simple basis for the cap are possible: for example, offsets might be allowed against sequestration using methods like scrubbers or biochar. But abuse of the 'paying someone else to cut their emissions' type of offset has to stop).

The Share

Next, the Share. In Figure 3, the curved arrows represent flows of money. The fossil fuel suppliers must buy the permits to cover the CO_2 content of the fuel they supply. They will seek to recoup this cost by building it into the price of fuel. This mark-up then flows through the economy like a carbon tax, making carbon-intensive goods cost more. There is thus a flow of money from the end-consumers to the fossil fuel companies, represented by the upper curved arrow in Figure 3. This sounds like bad news for the consumer. But wait – the fossil fuel suppliers paid for their permits, so where did that money go? The trick is to share this money out, back to the people (the lower curved arrow in Figure 3), which compensates for the price rises.

Figure 3

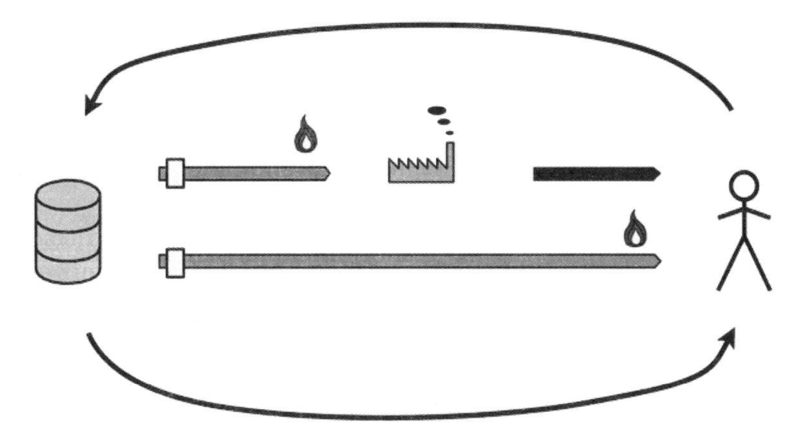

There are two possible mechanisms for returning ('recycling' or 'rebating') the money to the population. In the version called Cap & Dividend in the USA [2], permits are auctioned and the auction revenue distributed to the citizens on an equal per capita basis. Alternatively, under 'classic' C&S [1] each adult receives free of charge (say, monthly or annually) a certificate for his or her share of the cap – that is, of the country's carbon footprint. These certificates are then sold to the fossil fuel suppliers (through market intermediaries such as banks) and become the permits. Under 'classic' C&S people thus receive certificates instead of money, so that if they should wish to, they can retain (and destroy) a portion of their certificates – and thus are able to reduce the country's carbon footprint by that amount.

(Again, elaborations are possible. We'll consider these later).

That's C&S in a nutshell. It's simple and transparent; fair; cheap and fast to implement; positive and empowering; effective and efficient. The same principle could also apply to other greenhouse gases (see Chapter 8 by Richard Douthwaite) – in fact any other common resource such as a fishery could be incorporated: it is easy to maintain a cap using permits, auction these permits and distribute shares of the revenue to the population. This is a resonance with emerging so-called 'commons thinking' (Chapter 2).

Of course, a framework such as C&S is not the answer to everything: it is a complement to, not a substitute for, measures such as technology standards, tax regimes (e.g. support for renewables), education, and efforts to envisage and communicate a low-carbon future as a desirable one. It will not be sufficient to put the framework in place and 'let people get on with it'. But it is the framework which ensures that the numerical target set by the cap is met.

Winners and Losers

The payments to people compensate them for price rises, that is, the two curved arrows in Figure 3 balance. But this is only true on average. If you have a lower carbon footprint than the national average (remember that we are still talking about a scheme for a single country), then you will come out ahead: your payments from C&S will more than compensate for any price rises. On the other hand, people with higher than average carbon footprints will be worse off.

This means that there will be many more winners than losers. Why? For the same reason that there are more people on below-average incomes than on above-average incomes. Figure 4 illustrates this effect for an imaginary country consisting of only 10 people with CO_2 emissions as shown (in units of tonnes of CO_2). The total of all emissions is 15 tonnes and so the average is 1.5 tonnes.

Figure 4

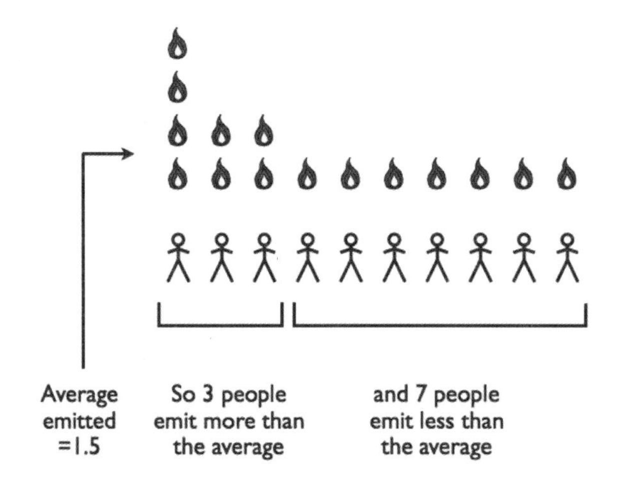

Thus a majority *would* be in favour of maintaining a tight cap, since they gain financially. This is a force to counterbalance the vested interests who would push for a cap to be relaxed or abandoned, and this counterbalance gives a certain political robustness to C&S in the face of shocks and political events. This robustness is necessary for a system which will need to survive for decades.

(Consider, by contrast, carbon taxes. A carbon tax is equivalent to an upstream cap if the tax level is set high enough. But the money disappears into general taxation, and so taxes are unpopular. So it is much less likely that the tax level would be set high enough. It might be a different story if carbon tax revenues were to be rebated directly to the population, but this is rarely proposed).

Of course there are some non-tangible gains, even for the people who are losers in strictly financial terms. For example, since everyone knows that the problem is being addressed, the rich can counter criticism from environmentalists by responding, 'my emissions are all within the cap too, and I am now paying for them, so stop criticising!'

More fundamentally, all are winners in the sense of benefiting from a stable climate. Indeed this, rather than financial gain or loss, ought to be the main argument for schemes like C&S. We face a collective crisis, and solving it effectively, fairly and collectively is simply the right thing to do.

Money isn't everything

The original formulation of Cap & Share envisaged a simple distribution of the benefits – in the form of certificates or their monetary equivalent – with equal shares going to the adult population. But exactly how to implement the Share is open to many different interpretations.

To begin with, one may ask why the *adult* population – what about children? Arguments can be made for partial shares to children. (Arguments can also be made for other adjustments to simple equity, such as allocating extra to rural households in developed countries. But special-case pleading could go on indefinitely, and there is a lot to be said for the simple guideline of equity).

Some argue against a distribution of money to individuals anyway. Many of the things that need doing in order to adapt to climate change are collective. Individuals don't usually build defensive sea-walls, for example. The original proposals for Cap & Share [1] include a Transition Fund, where a proportion of funds (declining over time) is held back and used for adaptation purposes, to help the people most severely impacted by climate change. On the other hand, at least in many developed countries, the individual payments would be seen as a popular factor in securing agreement to the policy, as outlined in the previous section.

One suggestion neatly combines the idea of collective funds with the issue of children mentioned above. If shares are to the whole population, not just adults, then the children's portion could support a Children's Fund to be used collectively in adaptation work and green investment – in other words, to benefit the world which the children will inherit.

An even more fundamental question is whether money should be distributed at all. Firstly, money is poor compensation for the climate being wrecked. But that is not what the Share is about – it is about *changing* the current destructive pattern of economic activity, not licensing its continuation. Secondly, there are many communities around the world which function well in the absence of monetary exchange and where its introduction might be disruptive. This is a powerful point. Certainly these societies should share in some way, but perhaps at a community or societal rather than an individual level. The communities themselves would be the best people to decide. This issue is discussed further in Chapters 6 and 7.

Perhaps the main conclusion from this discussion is that there is no single arrangement that is best for all circumstances. Different arrangements would be appropriate in different countries – this is as it should be, and probably applies at sub-national levels too.

But we must get going. Any debate over policy details takes place against the backdrop of a planetary emergency, where urgent action is needed despite the fact that no system will ever be perfect. The power of the idea of 'Share' lies not in the details adopted by this or that group, but in the general appeal to equity and to a shared effort to solve an urgent common global problem.

Comparison with Downstream systems

Cap & Share is an upstream system. To many people, however, the 'obvious' mechanism is not Cap & Share but a version of cap and trade applied 'downstream' where the emissions take place. This downstream approach may be an obvious one, but it is also more complicated. This is the route explored by various governments in developed countries, notably in the European Union.

In an upstream system, as we have seen, indirect emissions and direct emissions are covered by the same system. In a downstream system, however, they are treated separately. Indirect emissions are controlled with an Emissions Trading System (ETS) for companies, whereas direct emissions are covered by some form of personal carbon trading (PCT) [3]. PCT is based on ideas of 'rationing' and typically involves giving an equal allowance to each adult citizen, after which each purchase of petrol, oil or gas is deducted from the allowance (typically using swipe card technology). As shown in Figure 5, an ETS and PCT each operate at the point of combustion (which means the ETS is really 'midstream').

Figure 5

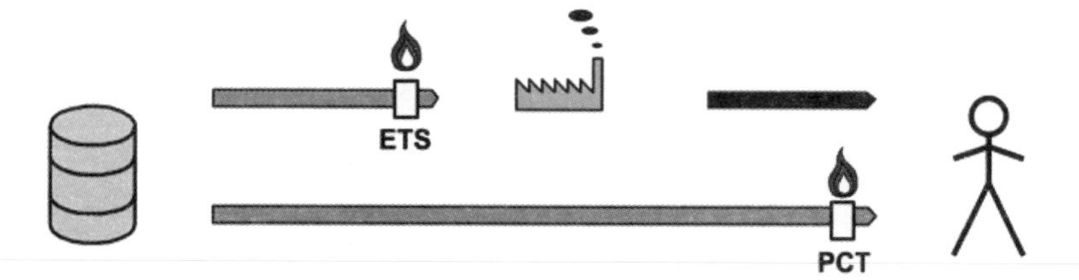

There is no reason why an ETS cannot have recycling of permit revenue to the population. To date, however, free allocation of permits has tended to predominate, generating windfall profits for ETS companies at the expense of the consumer. In the European Union, the EU ETS is already up and running, and has had its teething problems – lax caps through too many permits being issued, free allocation windfalls to large utility companies, partial coverage only of the economy, leaks through dubious 'CDM' projects and susceptibility to fraud. These shortcomings are now widely accepted and are being addressed in the next phase.

The downstream picture is complicated in practice by the treatment of electricity, measures necessary to avoid double counting, and the fact that the ETS leaves small companies outside the cap. These points are discussed below.

Hybrids

In practice, an ETS only includes large companies (although in the UK for instance, schemes like the Carbon Reduction Commitment extend the idea to medium-sized companies too). Hundreds of thousands of small ones are not in the ETS. So we have two sorts of company: those in the ETS and those that are not. In a downstream system the companies not in the ETS do not have emissions permits, so their carbon emissions are not captured by the cap. In Figure 6, notice that the middle line has no emissions permit straddling it. So the downstream PCT/ETS combination has incomplete coverage of CO_2 emissions.

Figure 6

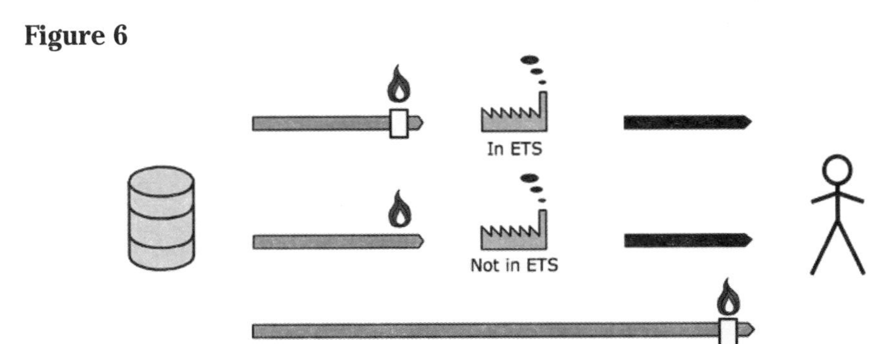

Supporters of the ETS concept want to extend it so that more and more companies are included – but this will always leave most small companies outside the scheme (and thus outside the cap). An ETS can never work properly with a downstream system like PCT.

An upstream system, on the other hand, can work with an ETS [4]. If for example we want to keep an existing ETS for political reasons, then introducing an upstream system is easy and can be made to dovetail with the existing ETS perfectly. To do this, start with a totally upstream system and simply allow the ETS companies to remain in the ETS, trading emissions permits. Fossil fuel suppliers must acquire permits for all fossil fuels as before, except that now any fossil fuels supplied to ETS companies are exempt. In Figure 7, notice that all three lines have an emissions permit straddling them, so this hybrid system has complete coverage of emissions. But we still have only a small number of companies involved (namely, the fossil fuel suppliers and the ETS companies), so the overall system remains simple and cheap to operate.

Figure 7

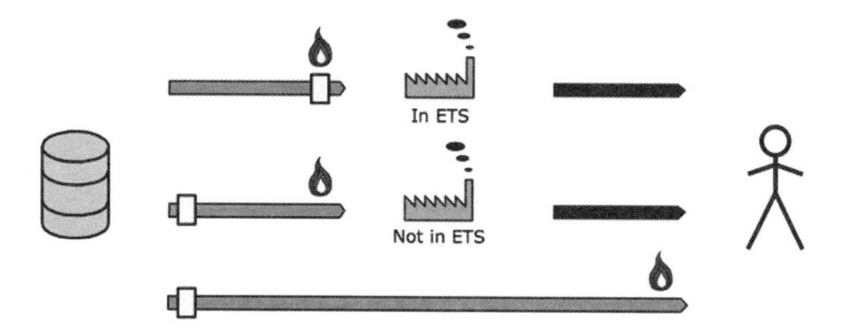

Electricity

Usually an ETS, and PCT as currently proposed, include electricity as well as fossil fuels. This makes sense in a downstream system, as electricity can be a substitute for gas in heating and cooking, and vice versa. So how does this affect things?

Using electricity causes indirect emissions, as the emissions are caused at the power station. We can depict this as in Figure 8 [5]. The three lines 'downstream' from the power station indicate the embedded or indirect emissions built into the electricity, and the flash symbols indicate electricity usage.

Figure 8

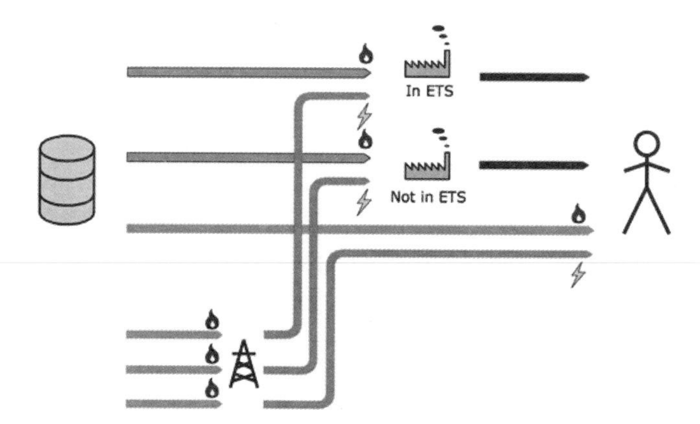

There are now 6 lines in the diagram leading from the fossil fuel supplier to the end consumer, instead of 3. But again all we need to do, in order to apply an emissions cap, is to decide where to put emissions permits to straddle each of these 6 lines. Cap & Share simply captures everything upstream (on the left hand side of the picture) as before. Once again, this is simple and gives complete coverage of emissions, as in Figure 9 (a). With a downstream system, we have the same problem as before because the ETS is only a partial system, but now there is another potential problem. Since the electricity generators are usually members of the ETS, there is a potential for double-counting: in Figure 9(b) two of the three lines which pass through the power station have two permits straddling them.

Figure 9

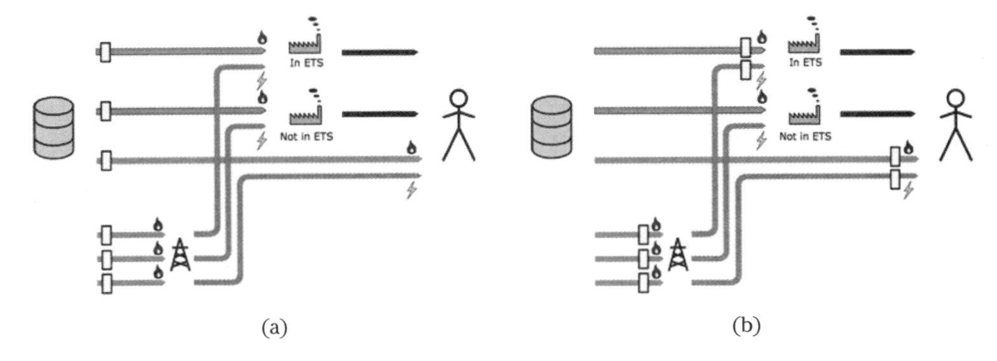

But once again an upstream system can embrace the existing ETS – with the permits as shown in Figure 10: it's easy to check that there is an emissions permit straddling each of the 6 lines. The rule is: fossil fuel suppliers need permits for all fossil fuels – but with fossil fuels supplied to ETS companies, or used to generate electricity supplied to ETS companies, exempt. Again this gives complete coverage, and yet only a small number of companies (the fossil fuel suppliers, the electricity generators and the ETS companies) are involved.

Figure 10

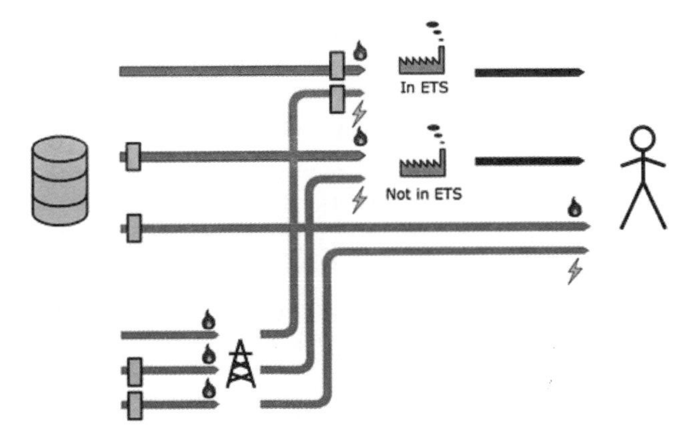

(Hybrids such as this are one way of introducing C&S 'gently' to allay fears and incorporate learning from other schemes. Other pathways are also possible: for example a government initially reluctant to impose a cap might introduce a carbon tax levied upstream; but this can easily morph into an upstream permit system with 'ceiling prices' and then (by raising the ceiling prices) into an upstream cap).

International versus Global

So far we've implicitly looked at a single country. What about the world as a whole? An obvious approach here is the *international* one, which seeks to add up and link together actions taken by the world's sovereign nations. In this approach a global cap is apportioned using a formula agreed by all, then each nation operates its own scheme (such as national C&S).

The apportionment formula between nations is of course a thorny question, as UNFCCC negotiations have shown. Suggestions include Contraction & Convergence (C&C) as promoted by the Global Commons Institute [6], under which national shares of a global emissions budget start at the current shares of global emissions and converge over time to equal per capita shares. Or there might be any number of ingenious elaborations on C&C performing a tricky balancing act of incentives. Or again, when the world recognises the extent of the emergency, we may be into Greenhouse Development Rights [7] territory – an approach which also explicitly addresses inequality *within* nations. Once again, the 'obvious' approach gets complicated very quickly.

In an ideal world, C&S would operate as a *global* scheme, a single policy for the planet considered as a whole. Figure 11 illustrates the difference between an international scheme and a global one. A global scheme needs a global institution to operate it, such as a Global Climate Commons Trust (Chapter 5), run perhaps by the UN. There would be a worldwide system of permits (which in this case would apply to digging up of fossil fuels only, since there are no 'imports' from other planets), with the resulting revenue returned to the (world) population. Global schemes thus bypass nations, except perhaps as a vehicle for transmitting the funds to their populations.

Global C&S is equivalent to national C&S in each nation with the national caps calculated on an equal per capita basis. So the eventual destination of many global and international frameworks would be the same. In fact, global C&S is just C&C with immediate convergence, and with 'the money going to the people'. (Once again, there are many possible elaborations. Some of the revenue could be kept back to fund collective projects, or to help specific countries with adaptation. Some proposals in fact, such as Kyoto2 [8], commandeer *all* the funds for such purposes).

Figure 11

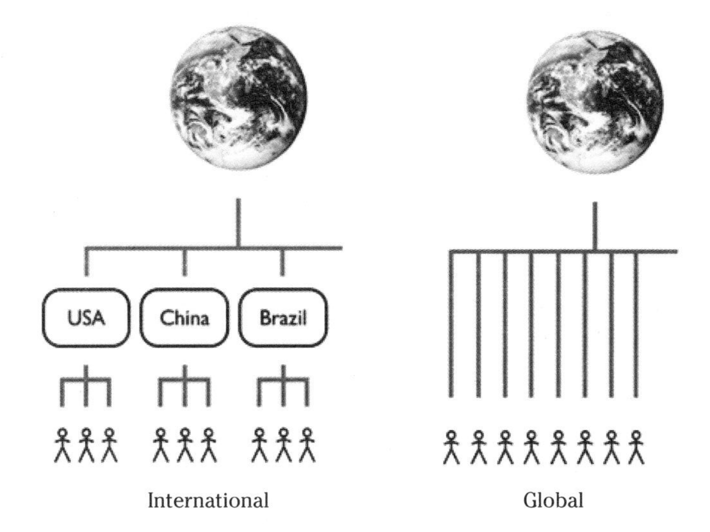

Global frameworks would require global institutions (and probably other things like monetary reform), and some regard this overruling of national sovereignty as hopelessly unrealistic – although others see climate change as a catalyst for wider reform. Global institutions would seem to be an obvious long term goal, but even international systems need global elements too: greenhouse gas concentrations are global entities and the cap must be set accordingly. 'Realism' is considered further in [9].

Winners and Losers Revisited

Extending the scope from a national scheme to a global one entails a change of perspective, from looking at the situation in one's nation to looking at one's situation in the world.

Suppose we look at another imaginary example, where the world has only 2 countries, each with 3 people with various carbon footprints, as shown in Figure 12.

Figure 12

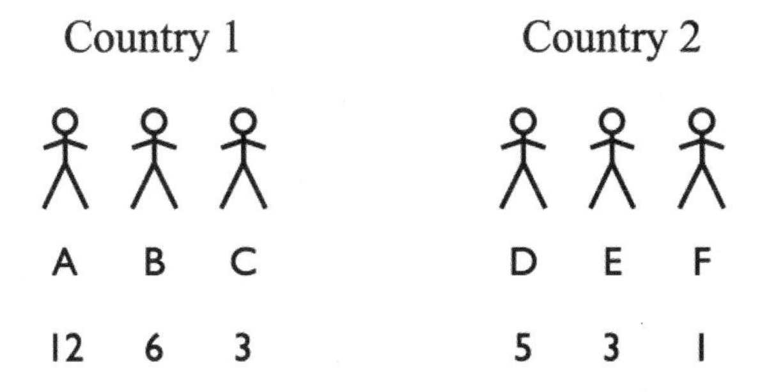

In Country 1 the average emitted is 7 tonnes. So B and C, being below average emitters, would gain if Country 1 adopted C&S (with C gaining more than B). These gains would increase as the cap in Country 1 tightens (for example, first moving to an overall reduction of 10% and then to a 20%, reduction). The same sort of effect happens in Country 2.

But this supposes that Country 1 and Country 2 act in isolation, which would have the effect of 'locking in' the inequality between the two countries. If, instead, there is any sort of convergence between countries, as discussed in the previous section, then the situation moves closer to one where the global average becomes the benchmark. Now the global average is 5 tonnes. C is still below this global average, but gains less than before. B, who was below average in Country 1, is above average in the world. Clearly people living in Country 1 gain now less than they did under separate C&S systems, and people in Country 2 gain more than they did.

This is an argument for building in convergence from the beginning before caps get too tight. Once C&S is well established in Country 1 in isolation (and achieving the desired effect of reducing Country 1's carbon footprint), moving from this to global C&S will be unpopular in Country 1. This issue has to be faced, in the same way that the issue of 'rich people are worse off under a cap' has to be faced in a national scheme – by pointing to the need for a collective solution to a collective problem, by appeals to equity and natural justice, by pointing out other non-monetary benefits, and by simply out-voting opponents. But underlying all this is the change of perspective mentioned above. B may cast envious glances at A, but is actually better off than the *whole population* of Country 2. It is natural to compare oneself with those closest to home, but increasingly, to solve global problems, we will have to recognise that we live in a single, interconnected world and think and act – and feel – accordingly.

Endnotes

1. Feasta (2008). *Cap and Share.* Dublin: Feasta. (www.feasta.org ; www.capandshare.org)
2. Barnes, Peter (2008). *Climate Solutions.* White River Junction, Vermont: Chelsea Green. (www.capanddividend.org)
3. Parag, Yael and Fawcett, Tina, eds. (2010). Climate Policy 10 (4), Special issue: *Personal Carbon Trading.* (www.climatepolicy.com)
4. Sorrell, Steve (2008). *Memorandum submitted to the Environmental Audit Committee.* In: Environmental Audit Committee (2008). *Personal Carbon Trading.* London: The Stationery Office, Ev 84-98. (www.parliament.uk)
5. Matthews, Laurence (2008). *Memorandum submitted to the Environmental Audit Committee.* In: Environmental Audit Committee (2008). Personal Carbon Trading. London: The Stationery Office, Ev 99-112. (www.parliament.uk)
6. Meyer, Aubrey (2000) *Contraction and Convergence.* Dartington: Green Books. (www.gci.org.uk)
7. Baer, P., Athanasiou, T. and Kartha, S. (2007). *The Right to Development in a Climate Constrained World: The Greenhouse Development Rights Framework.* Berlin: Heinrich Böll Foundation. (www.ecoequity.org)
8. Tickell, Oliver (2008). *Kyoto2.* London: Zed Books. (www.kyoto2.org)
9. Matthews, Laurence (2010). *Cap & Share: Simple is Beautiful.* In: Douthwaite, Richard and Fallon, Gillian, eds. (2010). *Fleeing Vesuvius.* Dublin: Feasta, 244-256. (www.feasta.org)

Chapter 4

Policy Packages

Nick Bardsley

A viable mechanism to reduce fossil fuel CO_2 emissions, such as Cap and Share,[1] is a necessary part of any coherent action plan to avoid catastrophic climate change, and end fossil fuel dependence.

There are many other aspects to the global economic and ecological crisis, though, which have to be dealt with through complementary and synergistic measures. Any element of a coherent package, including Cap and Share could, on its own, be counterproductive. This chapter explores some of the main issues that need to be addressed in tandem with capping fossil fuel CO_2.

It would take at least a book to do justice to any single one of the themes considered, however, and each has extensive further ramifications. For an attempt to flesh out the implications of ecologically-oriented measures across the policy domain, see Daly and Cobb (1994). What is attempted here is more modest, namely an exposition of some of the main issues interfacing with Cap and Share, and the related policy options featuring in current discussions. Cap and Share is our main measure to address the damage to the ecosystem caused by the global economy's unrestricted use of fossil fuels. It is a corrective to so-called 'market failure' at the heart of industrialised economies. Much of what follows considers other sources of market failure that need to be addressed at the same time for it to work. Working outwards from Cap and Share highlights some issues and not others. We do not discuss here several important issues related to the climate and energy crises which are pressing in their own right, including the measures necessary to reduce other greenhouse gases than CO_2.

---

[1] Or Cap and Dividend, or a carbon tax and dividend. "Cap and Share" in this chapter cab be read as a placeholder for any plausible CO_2 reduction mechanism.

Each section below outlines part of a climate change policy package that interfaces with Cap and Share. Although one element of this package, the Carbon Maintenance Fee, is seen as a global scheme, the focus is primarily on policy at the national level. This is notwithstanding the fact that much of what we suggest requires cooperation between nations that is currently lacking. The main aim is to clarify the nature of the relationships between the policy areas and to suggest some options.

1. Land - Based CO_2 Emissions: the Carbon Maintenance Fee

A substantial proportion of CO_2 emissions stems not from the burning of fossil fuels but from changes in land use,[2] such as deforestation and draining of peat bogs, and from carbon-depleting agricultural practices. CCSN (2010) estimate that around 30% of total greenhouse gas emissions are from land use, with around 1/3 of this attributable to CO_2. Various authors have suggested ways to curtail these, and means whereby the land might actually be used to draw down excess CO_2 by enhancing its natural 'carbon sink' function. The biggest single problem is with effective emissions from deforestation, which make up around 90% of CO_2 emissions from land use. There are currently an estimated 7500 $GtCO_2$ locked up in soils and vegetation (Stern 2006), which it is crucial to preserve. As explained in the chapter by Richard Douthwaite, however, it also seems necessary to increase this stock, to reduce atmospheric concentration of CO_2, since 350ppmv of CO_2 has already been exceeded. Significant threats to the current stock include continued deforestation and peatland degradation (CCSN 2010).

Approaches that have been suggested to tackle this problem include incorporating land use into the existing so-called 'Clean Development Mechanism' (CDM) instruments via credits from 'Reduced Emissions from Deforestation in Developing nations' (REDD). The CDM involves certification of projects which reduce emissions relative to a hypothetical scenario, the issuing of emissions credits corresponding to these reductions, and the trading of these credits for money on carbon markets. The premise is that it does not matter where emissions reductions occur, so long as they do occur. Those who are willing to pay more to emit can buy the rights to do so in return for emissions reductions elsewhere, so that emissions are reduced at least cost. The CDM has generated considerable trading activity. However, there are serious problems with the approach.

The CDM is widely believed to deliver few (if any) emissions reductions, at a high cost, to be extremely vulnerable to fraud and to generate perverse incentives. A key problem is that the assessment of emissions abatement relative to a hypothetical future does not imply emissions reduction in an absolute sense. The "reductions", that is, are only relative to "what would

[2] So-called 'LULUCF' emissions: Land Use, Land Use Change and Forestry.

have happened". Regarding perverse incentives, two examples are indicative. Many hydropower schemes have apparently been certified that would have taken place even without the CDM; and the production of HCFCs has been stimulated in order to generate HFCs, in order to earn credits from abating HFCs HFCs (Tickell 2008, p34-37).[3] Similarly there is a danger that the inclusion of reduced deforestation will create a perverse incentive to increase rates of deforestation in order that more credits could be earned by a given degree of restraint. One could also expect false claims of afforestation and reforestation to accompany industrial forestry or plantations, which have very different characteristics to naturally occurring forest.

To avoid such issues arising a far simpler approach, advocated by FEASTA, is to pay countries a fee for the carbon stored in their forests and soils, and for this to be assessed via auditing procedures including remote monitoring. This 'Carbon Maintenance Fee' (CMF) would thereby give nations an incentive to maintain their forests, to increase the carbon stored in plants generally, and to increase the carbon content of their soils.

The CMF would operate as a global fund overseen by a global climate trust. Countries could contribute to the fund's costs proportionally to their incomes. FEASTA propose a two-part, annual payment from the fund to each country. The first part of the payment would be based on the estimated mass of carbon in a country's soils and biomass in the course of the year. The second payment would be for any increase in the stock of carbon that had occurred during that year. There would be a corresponding financial penalty if the stock of carbon had decreased over the period. The scheme is illustrated in the Box below.

Aside from curtailing deforestation and peatland degradation, new activities that this could stimulate include greater use of organic agriculture, and the widespread production and application of biochar from crop wastes. Biochar is a form of charcoal formed by pyrolysis and charged with live organic matter such as compost. See the chapter by James Bruges in this volume and Bruges (2009).

Issues raised by the CMF include how countries might use the funds and how to finance it. Concerning the former, how it would be implemented within a country would be up to the national government. Incentive payments to farmers are one option. An example comes from Costa Rica, where forest cover increased from 22% in 1977 to 51% in 2005, thanks in part to a $45/ha afforestation incentive payment to farmers (Stern 2006). At present making payments conditional on soil carbon improvements on specific land holdings seems too demanding in terms of accuracy of measurement at high spatial resolution. Monitoring of agricultural activities and other land use practices,

---

[3] HCFCs are refrigerant gases, damaging the Ozone layer, controlled by the Montreal Protocol. HFCs are alternative refrigerants falling under the Kyoto Protocol, and extremely powerful greenhouse gases. HFC-23 is a byproduct of HCFC production.

to estimate carbon sequestration at a local level, seems more feasible (Byrne 2010). It would make sense for incentive payments to go to farmers rather than landowners. However, one problem would be that rents and land values could increase, reflecting the value of the new subsidy, providing windfall gains to landowners. To counteract this, a land value tax could be introduced, whereby landowners would pay either a regular fee or a percentage of the sale price of land.

A land value tax would tap windfall gains deriving from land ownership generally, and raise substantial revenues. Those revenues would also be available on a long term basis. This is appropriate because carbon sequestration will be a very long term process, extending beyond the Cap and Share period. The tax could therefore help nations finance their contribution to the carbon maintenance fee. Land value taxation is discussed further in section 7. An additional source could be revenue from an international 'Tobin tax.' A Tobin tax is a small tax on international financial transactions, which would also help to curb speculative financial flows and capital mobility, the magnitude and volatility of which are widely recognised as undesirable features of the current system.

As Tickell (2008, 46-47) argues, however, incentive schemes to governments to halt deforestation risk violating land rights of indigenous peoples and / or existing commons management arrangements. Payments should therefore be conditional on respecting such rights and practices. Furthermore, incentive payments may not always be the best approach at ground level. For example, it may often be the case that strict protection of existing land use rights and commons management arrangements from competing interests provides the best means of preserving and enhancing forest carbon stocks, rather than incentives payments to famers, foresters and so on. Monetary incentives plausibly crowd out intrinsic motivations. This is evidenced in Titmuss's (1970) classic comparison of blood donation in the UK and USA, for example, and in studies in behavioural economics (Gneezy and Rustichini 2000; Bowles 2008). It may therefore be counterproductive to introduce payments in contexts where protection can be afforded without them.

The Carbon Maintenance Fee

To illustrate the proposed incentive scheme, we present calculations based on Byrne (2010). The basic carbon maintenance fee would generate a small payment per ton of CO_2. A 10¢ per ton fee would amount to an initial $204bn set of transfer payments since there are currently an estimated 2040 Gt of carbon stored in the world's vegetation and soils. Transfer payments should not be considered a cost to the world since total world income is not affected. The transfer is received in return for a service provided to all peoples. How the funds are used would be left to individual countries to decide for themselves.

An additional transfer is proposed to provide strong incentives for carbon sequestration and penalties for CO_2 emissions. Sequestration would be rewarded and emissions penalised, at the CO_2 price (determined by Cap and Share). Consider deforestation. It is estimated that a hectare of forest contains an estimated 1000 tons of CO_2. If income from crops and sale of timber amounts to an estimated value of just over $2000 / ha (the estimate from Stern, 2006), any CO_2 price above $2 per ton would make it unprofitable to deforest, even ignoring income foregone by additional carbon sequestration. The penalty could be levied by withholding the basic maintenance fee. Given that realistic carbon prices would be well in excess of $25 per ton CO_2 there would be very strong incentives operating to end deforestation.

Standing forest sequesters additional CO_2 each year, so payment of the CO_2 price for additions to the stock would constitute regular income for nations maintaining or growing their forests. A recent study in Africa estimates the sequestration rate there as 2.2 tons CO_2/ha. This implies African forest sequestration of 1.2GtCO_2/yr, so an income of $30bn at a CO_2 price of $25/t$CO_2$, or over 10% of total Sub-Saharan African export earnings in 2007. This constitutes an additional (transfer) funding requirement for the scheme, which could operate on a national income tax basis.

2. Land Use Controls and Food versus Fuel

One problem with implementing Cap and Share on its own would be that, as fossil fuel use is curtailed and its cost rises, there will be an incentive to substitute into other forms of energy. This is part of the point of the policy, but alternative forms of energy supply are not equally desirable. Perhaps the most evident problem arises from substitution into agro-biofuels, that is, energy produced from crops. As fossil fuel prices respond to the cap, farmers may divert agricultural land to the production of oil crops, such as jatropha or oilseed rape, or sell food crops such as maize to biofuel refiners instead of into food distribution. There is also a knock-on problem of indirect land use change resulting from this, as more land will then have to be cleared to produce food to cover the shortfall. This may happen through the free market in response to higher food prices. It would probably result in increased deforestation, biodiversity loss, infringements of indigenous people's land rights and so on – a familiar set of adverse development outcomes.

Figure 1

Tortilla Protest in Mexico, 2007. Photograph by 'Lizdinovella.' The banner reads "Without Corn, there is No Country."
Reproduced under creative commons 2.0 license; http://www.flickr.com/photos/lizdinovella/2741801519/

The biofuels industry, spurred by biofuels directives and subsidies in the developed world, has already been implicated in food shortages. In 2007-8 there were widespread disturbances, sometimes violent, in Less Industrialised Countries (LICs) as the cost of living surged for many of the world's poorest people. In Mexico, for example, protestors took to the streets in response to the rise in price of basic foodstuffs, such as the tortilla. This resulted from biorefiners, principally in the United States, buying up the maize crop for conversion into ethanol. By February 2007, the price of tortillas, a staple food and key calorie source for poorer Mexicans, had risen by over 400%. A leaked internal World Bank report (Mitchell 2008) subsequently attributed approximately 75% of the price increases to the expanding bioethanol market. Other reports give lower estimates, citing other factors including crop failures elsewhere and speculation. However, part of the reasoning of the World Bank report was that speculative activity was amplifying the price increase initiated by fundamental factors.

In mid-2011, the UN FAO Food Price Index showed food prices to be higher in real terms than in 2007-8, at more than double their 2002-4 average level. This has triggered further unrest, and appears to be one factor behind the 'Arab spring' uprisings.

Biofuels are discussed further in section 5. Whether there is any defensible role for them in the future world economy is controversial, and depends crucially on their land use requirements and energetic characteristics such as Energy Return On (Energy) Investment. Here we ask what can be done to counter this 'food versus fuels' problem in the absence of a global ban. The Carbon Maintenance Fee would offer inadequate protection here because it only operates on the land use change aspect of the problem. That is, it does nothing to prevent crops being sold to feed cars rather than people. Also, regarding the land use aspect, the CMF would have to be set at a high enough level to discourage deforestation for crop production. There is no guarantee, however, that those administering the scheme would get this right, or be able to fund a high enough level of CMF.

To prevent adverse land use change, spatial planning policies should decree that agricultural land be designated for food production, and forests protected. To curtail the sale of food crops for refining, certification schemes have also been proposed to ensure that biofuels have been produced using non-food crops on land unsuited for food production. The effectiveness of such measures is likely to depend on the amount of resources that a country can devote to enforcement, and this may decline as diminished energy availability sets in. It will also vary inversely with the incentives to cheat or defy the system, which will be greater the higher the carbon price. Thus it is possible that more direct powers of the state may have to be invoked, such as banning exports of fuel foodstuffs, nationalising biorefining (if and where it can be a viable part of the energy mix) or actually taking over functions of crop purchasing and retailing to prevent their diversion to biofuels internally.

It is in LICs where the food versus fuel problem is most acute, since a far higher proportion of income is spent on food there than in the industrialised world. Many LICs have recently been involved in deals which seem certain to reduce their food security though. Corporations typically from richer countries have, amidst the ongoing financial crisis, been buying or leasing large tracts of agricultural land in LICs to secure supplies either of food or biofuel (GRAIN 2008, 2009). In addition to violating existing land use rights, with consequent social unrest, this trend seems likely to lead ultimately to increased geopolitical instability (that is, food wars). Further comments on this situation are given in section 8.

Advocates of 'second generation' biofuels, notably lignocellulosic fuel produced from fast-growing grasses and trees such as switchgrass and miscanthus, have also advocated a lower degree of meat consumption to free-up land. See for example, in a UK context, CAT (2010). However, it is questionable whether biofuels make sense energetically. This matter is discussed under energy policy in section 5. The most critical biofuel analysts imply that the food versus fuels problem is entirely an artefact of misguided subsidies and policy directives.

3 Agricultural Policy

Reduced fossil fuel availability implies huge changes for agriculture, necessitating an extensive program of support and reform. This is because of the extent of fossil fuel use for synthetic fertilisers, pesticides, irrigation, traction and transport of produce. According to professor Albert Bartlett, the use of oil in industrialised farming systems is so pervasive that "Modern agriculture is the use of land to convert petroleum into food" (Bartlett 1978, p880).[4]

There are implications for both the scale of agricultural activities and the techniques employed. Petroleum-based agriculture has produced huge, minimally employing, mechanised farms serving remote markets via monocultures requiring huge synthetic resource inputs. With less energy throughput, production needs to be closer to local self-sufficiency on a bioregional basis. So food supply has to become less based on international trade and specialisation. It also seems clear that labour must substitute for capital, that organic methods should displace synthetic pesticides, herbicides and fertilisers, and that alternative means of traction will need to be deployed. On-farm energy needs may be met partly by methane supplied through anaerobic digestion of farm waste and manure.

The challenges are exacerbated by the current widespread use of 'F1 hybrids' and the increasing use of GMOs. The former, and in practice often the latter, preclude seed saving and the latter are frequently tied to intensive use of inputs and often specific chemical packages. Deprived of these they offer less disease resistance and general hardiness than the inherited seedstock, 'landraces', that they have displaced, often to extinction. Thus, the revival of traditional varieties, and seed saving under different local conditions to encourage genetic diversity, seems vitally important.

---

[4] This section draws freely on Daly and Cobb (1994, ch. 14).

Organic agriculture should be understood as a set of positive practices, rather than simply coping with an absence of chemicals. The starting point is consideration of what generates healthy soils, crops and livestock, rather than how to intervene when things go wrong. For examples of techniques that follow, green and animal manures and leguminous crops help provide fertilisation, and integrated pest and water management reduce external input requirements, and biodiverse pastures provide a varied diet for cattle and reduce veterinary requirements. Such methods often require reduced spatial scale. For example, bugs accommodated in hedgerows, which eat pests, have a limited range (circa 200m), so vast field sizes become counterproductive.

The changes required cannot happen overnight, so countries should be gearing up now for what is in store. It is worth noting that it is not impossible for lower input and more ecologically oriented agriculture to succeed even in the current system, as the organic farming sector demonstrates. One can also point to the survival of the Amish farming communities in the U.S., who deliberately minimise their use of exosomatic energy. However, these examples are on the periphery of current practice. Policy measures that could be enacted now to help the necessary transition include an end to subsidies structured to favour large scale agribusiness; these also inflate land prices. In the same vein, pollution taxes would disfavour practices which rely on mechanisation and chemicals instead of stewardship of the land. Land Value Tax would help by increasing the supply of agricultural land, lowering its price and enabling more people to farm thanks to lower start-up costs.

Also needed would be a support package involving 'extension work' in relation to organic agriculture. This is the communication of innovative practices to producers, and provision of support for their use. Farmers are often said to be conservative in relation to alternative approaches to food production, in particular ecologically-oriented approaches.[5] A reluctance to experiment is understandable in view of both the variability of nature and competitive pressure in food markets, where market power is very concentrated amongst large food retailers. The extension work should accompany significant reallocation of research resources away from industrialised agribusiness to develop expertise and technology for agro-ecological methods. The need for greater development and application of such techniques has also been argued on independent grounds. These include adaptation to climate change and the inappropriateness of corporate agribusiness practices in an LIC context (IAASTD 2008).

---

5 Mike Bryant, University of Reading, personal communication..

A case study of what may lie ahead agriculturally is provided by Cuba's experience in the 'special period' of the early 1990s. Cuba's relatively industrialised economy, and heavily industrialised agricultural sector, had developed with the help of Soviet oil. After the collapse of the Soviet Union, oil was only available to Cuba through the world market at dollar prices. This led to a severe fuel shortage, a situation exacerbated by extensive trade sanctions against the country.

An inspiring, if somewhat rose-tinted, rendition of Cuba's experiences is given in the film 'The Power of Community,' which emphasises the extensive use made of organic agriculture and permaculture techniques, particularly in urban market gardens. A more comprehensive and contrasting picture is given by Wright (2009), who reports that the adoption of organic methods was much less extensive in rural areas, and that as a whole the shift is better described as one towards low-input farming than towards intentionally organic methods. The urban organic sector is estimated by Wright to have contributed around 5% of total food supplies in this period, mostly fruit and vegetables for local consumption.

The immediate aftermath of the oil crisis sparked a dramatic fall in agricultural productivity and shortages of food. As a result of this, food items were rationed that had been freely available previously. Oil was prioritised for electricity production, so far less was available for agricultural uses than previously, and fossil fuel-derived fertilisers and agrochemicals were more difficult to obtain. To boost production levels a greater proportion of the workforce had to be recruited into agriculture, animal traction reintroduced, organic methods introduced and biological substitutes developed for pest control. A conflict remained between boosting productivity and sustainability.

Barriers to the take up of new methods apparently included a degree of conservatism in the rural community, with new, urban producers being more open-minded regarding alternative techniques. Wright estimates on the basis of surveys that there were no intentionally and wholly organic farms in rural Cuba, with 83% of farmers wishing to use more agrochemical inputs, though there seemed to be little desire for fully-fledged return to high input practices. Wright also reports institutional resistance to innovation from agencies and departments that had developed expertise under the old input intensive system.

A striking feature of the Cuban experience is the variety of agricultural organisations that were developed, including different kinds of cooperative enterprise, private farms and smallholding, coexisting with large state farms, and varied distribution channels, partly it seems, in an effort to find what works. Many farms were downsized and use rights to myriad parcels of land were handed out in 'usufrucht,' meaning the right to use and profit from an area of land without actually owning it or having to rent it.

Many countries share a pattern of agricultural development similar to Cuba's prior to the special period. That is, they exhibit increasing farm sizes, a high degree of mechanisation and use of petroleum derived-inputs, and the progressive concentration of food retail. It seems likely that this pattern will have to reverse as the same inputs become scarcer under the cap. Land reform and the breaking of monopolistic control of the food system may also be necessary. This would facilitate a 'return to the land,' and creative experimentation with substitutes for conventional techniques in different growing conditions. There may be a need for some nationalisation of land, the granting of usufrucht rights and the imposition of a land value tax to increase the farmland available for rent.

Finally, Wright suggests that the "main lesson" for the world from the Cuban episode is the importance of planning for the possibility of food rationing and direct government intervention in supply chains, in response to shortages. This may sound extreme, even absurd, to those whose working lives have been defined by the post-Soviet era and (in the North) an age of apparent plenty under minimal government intervention. But a rapid decarbonisation of the world's economies is an extreme prospect implying dramatic adjustments.

We mentioned above in the context of the Carbon Maintenance Fee that there are land use practices that may significantly enhance carbon sequestration. Here, we note the synergy between such measures and agriculture. There is evidence that established methods of organic agriculture increase soil carbon content over time (Soil Association, 2009). Innovative restorative grazing practices, pioneered by Alan Savory in the plains of South Africa, suggest that properly rotated free-ranging livestock herds could also sequester carbon (CCSN 2010). This involves continuous moving of the livestock, in imitation of naturally free-ranging herds kept in motion by predators, with the effect that the grasses put down a larger mass of roots. This appears to have numerous effects serving both to regenerate the local ecosystem and incorporate carbon into the soil. Biochar, as discussed in James Bruges' chapter, appears to have an important role to play in both carbon sequestration and enhancing soil fertility, with additional potential benefits from a biofuel byproduct.

4. Monetary Reform

> "The modern banking system manufactures money out of nothing. The process is perhaps the most astounding piece of sleight of hand that was ever invented. Banking was conceived in iniquity and born in sin. Bankers own the earth. Take it away from them, but leave them with the power to create credit, and with the stroke of a pen they will create enough money to buy it back again. ... If you want to be the slaves of the bankers, and pay the costs of your own slavery, then let the banks create money."

The quotation is attributed to Sir Josiah Stamp, director of the Bank of England in the late 1920s, though this is not verified. The sentiments it expresses are defensible, though they may sound extreme. That banks create credit out of nothing, with minimal backing in deposits, is a fact recognised by orthodox economics in its basic textbooks. The matter is generally quickly passed over, however, and contradicted elsewhere in the corpus where banks are characterised as 'financial intermediaries', redirecting funds from lenders to borrowers. The credit banks create, which also constitutes debt, makes up the vast majority of money supplies in developed economies, typically 95% or more, with notes and coin making up the remainder. Virtually all money exists, then, only because interest-bearing debt has been incurred.[6]

There are strong reasons for considering the money system in connection with climate change. GDP, a measure of the level of economic activity, is, unsurprisingly, strongly related to greenhouse gas emissions. Growth is commonly defined as real GDP growth. And real GDP is generally dependent on the quantity and sectoral allocation of credit (Werner 2004). It follows that emissions growth is dependent on credit growth. Credit growth seems to be an intrinsic feature of the modern 'fractional reserve' banking system, and

[6] This section draws freely on Werner (2004).

may even be necessary to stop it from collapsing. Modern banking therefore seems to have emissions growth written into it.

Various positive feedbacks can be identified in the credit creation system. Firstly, the interest paid on loans provides profits to banks, and these profits underpin further loans. Secondly, the current era of deregulation has resulted in minimal and apparently ineffective reserve and capital requirements across much of the global economy. £1 received in deposits may therefore enable more than £1 in additional loans directly, by the same bank. Thirdly, growing credit creation increases economic activity, which then strengthens the confidence of banks to issue loans. Continuous credit expansion is evident in the graph below, showing notes and coin (M0), deposits in all kinds of bank account (M4) and the stock of debt owed to the banks (M4L) in the UK. That banks expect such an expansionary process is evident in the compound interest normally offered to savers, which cuts the savers in on bank profits.

Figure 2

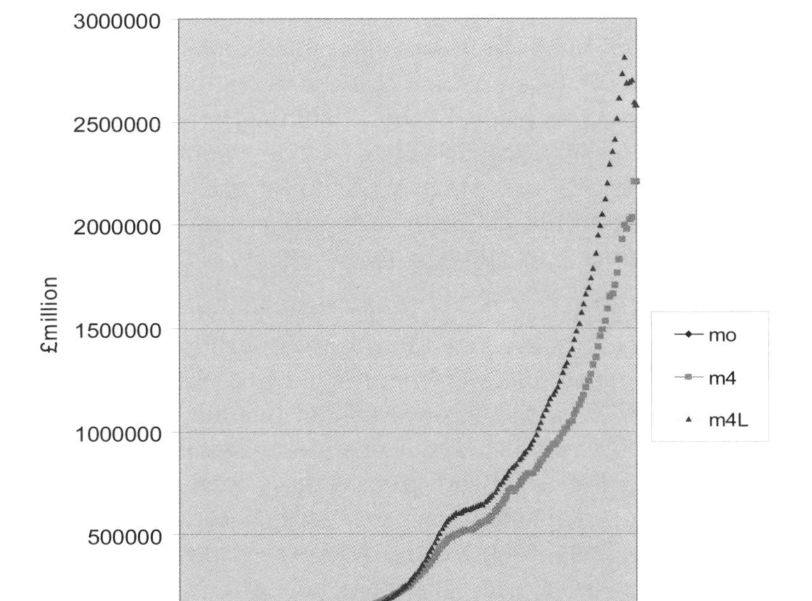

Continuous expansion of the UK money stock (M4) and lending stock (M4L), 1963-2010. M4L fell for the first time ever in publicly-available data in March 2009. Only a around 3% of the UK money stock took the form of notes and coin in April 2006, at which point M0 ceased to be published.

Source: Bank of England, online database, accessed July 2010.

For *real* rates of interest to be earned, that is, above the rate of inflation, there also has to be greater provision of goods and services. In practice, expanded credit is issued in order to facilitate both production and trades in existing assets. The price of existing assets, such as most stocks and shares and real estate, is sensitive to the availability of credit. Continuous credit expansion therefore implies a mix of growth and asset price inflation. Before the Thatcher / Reagan era of deregulation, attempts to control this mix had been made to ensure enough growth, via steers on the types of activities that bank lending could support. With deregulation, there has in contrast been a proliferation of house price bubbles across Europe, the USA and much of Asia. The tendency of asset prices to be bid up by a growing volume of credit gives rise to an additional positive feedback in the upswing of the business cycle, as analysed by Minsky (1982), since assets serve as collateral on loans.

Conversely, it is not clear how the interest owing on loans can be paid if money incomes are not growing. For only the principal on loans is created when they are extended. For real interest to be paid, either economic activity must expand or redistribution to creditors takes place. In a fossil fuel economy, production can be expanded through a combination of increased exploitation of primary energy sources and energy efficiency gains. But in a world powered by renewables it is doubtful whether comparable expansion is feasible. Energy from a stock of fuels can be converted at a chosen rate, whereas a flow from a renewable source requires a specific duration. Energy analysts consequently struggle to devise plausible scenarios in which renewables replace fossil fuels at current rates of energy use. Mackay (2009) for example, fails to reconcile current energy use for the UK with renewable energy under deliberately optimistic assumptions. The energetic requirements of continued exponential growth are considerably more exacting.

We believe, therefore, that a transition away from fossil fuels needs to be accompanied by a transition away from the current money system. If growth is no longer possible then real interest cannot be paid on a sustained basis. There are additional reasons in favour of a new system. The financial sector, which produces only the credit with which others conduct real business, is in effect a drain on the real economy because of the interest it charges. Interest supports a "rentier class" who live by harvesting interest payments, often supported by inflated asset prices including house and share prices. Although that class is participated in to some extent by anyone holding interest-bearing savings, financial sector growth has contributed to increasing inequality. Under a declining carbon cap, greater equality would need to substitute for growth as a source of improved social welfare. Finance would also need to be channelled away from speculation towards productive uses.

According to data from the OECD, shown in Figure 3 below, the financial sector now ranges from 25% – 33% of GDP in the USA, Japan, UK, France and Germany and is now often larger than all the productive sectors put together. This figure understates the significance of finance in the economy, since trades in existing assets are not included in the calculation of GDP. The sector's share in corporate profits rose from 10% in the USA in the early 1990s to 40% in 2008 (Gudmundssen 2010). Whilst there is a pressing need for a dramatic realignment of economic activity along ecologically rational lines, an ever greater proportion of economic resources has actually been allocated to the financial sector.

One alternative to the current system, which would allow a deliberate reorientation, is one in which money is spent into existence debt free, rather than borrowed from private banks. Governments, both local and national, are (in principle) in a position to do this because they can decree that currency they issue be accepted for taxation payments. This ensures a demand for the money amongst all who have to pay taxes, and therefore further secures its acceptability as a means of exchange for other transactions. Numerous thinkers have set out proposals based on this possibility, and there are several historical examples of debt free money, spent into existence by government. A useful guide is Rowbotham (1998). Fewer attempts have been made, though, to address the interface between monetary reform and the ecological – energetic crisis.[7]

Rowbotham offers proposals based around a citizen's income, based on the suggestions of C.H. Douglas, founder of the 'social credit' movement. Each adult citizen would receive a basic income from the state, spent into existence or recycled from taxation as necessary, which would enable much of the means-tested welfare system to be done away with. Others have proposed that the government becomes a more influential player in the economy, for example by directly financing large capital expenditures. With the cooperation of the central banks, the proportion of government-created money could be set to increase over time, displacing that created by the private banking sector.

---

7 At the time of writing in the UK, for example, there is the 'Positive Money' campaign. This argues for 100% reserve banking, under which banks can only lend out money that their customers agree not to withdraw whilst it is out on loan. This would make the money system more stable and manageable, but would not address its need to grow indefinitely if loans and savings accounts still bear interest.

Figure 3

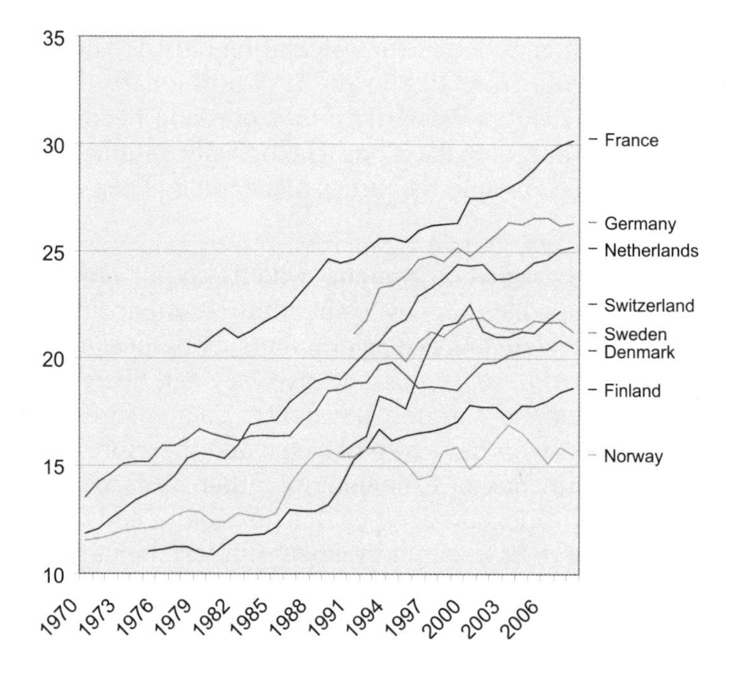

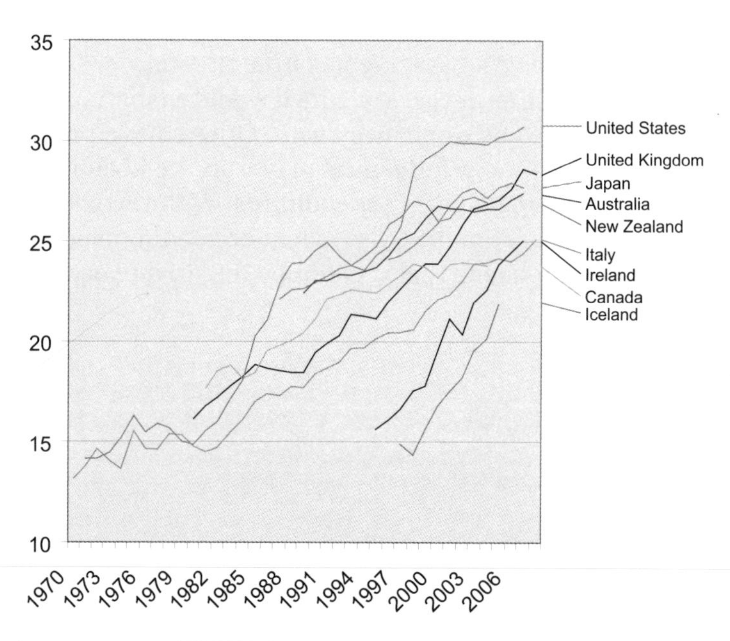

The Size of the financial sector 1970-2008. Source: OECD Stat online database; from series B1GJ_K: Financial intermediation, real estate, renting and business activities. Accessed August 2010.

Integrating such proposals into a climate change mitigation package seems a very promising route. Initially, for example, expenditures of government-issued credit could be used on such expenditures as purchases of renewable energy infrastructure and a comprehensive thermal upgrade of housing stocks, on a zone-by-zone basis. Citizens' income payments could perhaps be phased in, as people's income from sales of Pollution Allocation Permits declined, as fossil fuels are phased out. Finally, in line with the likely relocalisation of the global economy, regional currencies and associated credit creation powers for local authorities may have an increasing role to play (Douthwaite 2010).

At the time of writing the credit creation feedback loop seems to be operating in reverse in many countries. The crisis in the Eurozone is dominating the news agenda. However, many of the recent problems originate in domestic banking developments that are not driven by the specific problems of the Eurozone. Iceland, whose banking system collapsed in 2008 following a massive credit bubble, was not a member, and its problems preceded the Eurozone crisis. The United States is experiencing a credit crunch in large part attributable to a declining housing market, again following very rapid credit expansion and house price inflation. Essentially the same process had taken place in Scandinavian countries in the early 1990s, and in Japan in the late 1990s. The Eurozone situation does illustrate the importance of monetary autonomy in responding to a crisis, however. This can be seen from comparing the cases of Iceland and Ireland. Iceland's currency was massively devalued in the wake of the crisis. Whilst this caused considerable hardship in the short term, as imported products increased sharply in price, it has enabled a degree of economic recovery. Ireland, in contrast, cannot devalue because it no longer has an independent currency.

The current crisis may have been triggered by higher energy costs, which may be attributable to peak oil. If so, the value of money must fall in relation to energy, and this may be occurring via the withdrawal of money from circulation (Douthwaite 2010). The long term danger posed by the resultant situation is that finance will not be available to support measures to achieve a low carbon economy whilst there is still a relative abundance of cheap energy to pursue this. The issue will also, probably, disappear from public awareness, amidst hardships such as unemployment and bank seizure of collateral assets, which are more visible and immediately pressing. Both to deal with the recession in a more humane manner and to support green investment, monetary autonomy seems crucial. This would be maximised if the right to issue currency were reclaimed from private interests and made to operate in the public interest. Inadequate monetary autonomy extends beyond the Eurozone. All signatories to the Maastricht treaty, for example, have pledged not to use central bank credit to fund government deficits.

As part of the response to the current debt crisis we propose an 'ecological debt jubilee' (see the box below) which would allocate a substantial sum of money, in voucher form to direct its use, to each citizen, whether or not they are indebted. In addition to relieving an excessive debt burden, this would increase the proportion of debt-free money in circulation.

An Ecological Debt Jubilee

A debt cancellation is the original meaning of the word "jubilee". Jubilees appear to have been common in the ancient world, for example taking place every 50 years in Mesopotamia around 2000BC. A biblical description of a jubilee occurs in Leviticus 25. In the recent credit / debt expansion, debts have outpaced income growth. This, coupled with increases in energy prices and associated inflation, has rendered continued debt service intolerable for an increasing proportion of the population. Some form of debt write-off therefore seems inevitable. This could take place in an orderly or disorderly fashion, with disastrous consequences for the populace under the latter. We propose a scheme which would preserve the functionality of banking whilst it is being re-engineered as a very different system, in which the power to create money is no longer privatised. This re-engineering is necessary lest we return after a short interval to the same situation, the eventuality foretold in the quotation at the start of this section.

A simple write off of debt would have the consequence that banks would incur massive losses, since debts are accounted as assets, but liabilities would be left intact. In that case banks' incomes would be insufficient to cover withdrawals, with a consequent run on the banks and breakdown of the money supply. We suggest as an alternative that vouchers, worth a substantial proportion of average personal debt, be distributed to each adult citizen, which can be used by borrowers for debt reduction. The vouchers can be exchanged by the banks for new credit issued into their accounts at the bank of England.

Since debt relief would be unfair, absent compensation, to those who are not in debt, they would be allowed to use the vouchers for other, specified, purposes. Given the climate / energy emergency, we propose that householders would be allowed to use them for purchases of energy efficiency improvements, solar panels, solar thermal systems, heat pumps and so on, from existing approved suppliers. This would prevent 'cowboys' cashing in on a boom.

Renters, or others who do not need such improvements, would be able to purchase renewable energy bonds, which can be redeemed for kWh in energy bills at a future date, purchased at current prices. The proceeds from these bonds would be used to fund investment in renewable energy technologies. The energy companies issuing the bonds could exchange the vouchers for (debt-free) credit issued by the central bank, via their own banks acting as intermediaries. Since the price of energy can be expected to continue to rise, this constitutes a good investment opportunity for the bondholders.

The cancellation of debts in this manner would result in the circulation of money that never gets retired as debt is paid down – that is, debt-free money. Its effect on the money supply is net positive. This is what is required in a situation where money is disappearing as debts are paid down. The citizen debt jubilee could form part of a larger issue of debt-free money disbursed to relieve states of their excessive debts, as proposed by Richard Douthwaite for the Eurozone.[8]

5. Energy Policy

Synergy with a Carbon Cap

Cap and Share has a clear synergy with energy policy, in particular policies to promote energy efficiency and renewable energy technologies. These are needed to maintain tolerable living standards, if at significantly reduced levels of consumption in the North, whilst moving away from fossil fuels. Improving energy efficiency means increasing the amount of useful work that a given amount of energy performs for us. There is broad support for energy efficiency, since it is generally seen as both profitable and environmentally friendly.

Similarly, most, environmentalists are committed to the development and diffusion of modern renewable energy technologies. This position is not uncontroversial, however. See for example, the recent exchange of letters between Paul Kingsnorth and George Monbiot.[9] Kingsnorth argues that it is

8 http://www.feasta.org/wp-content/uploads/2011/07/Deficit_easing_RD.pdf
9 http://www.guardian.co.uk/commentisfree/cif-green/2009/aug/17/environment-climate-change

futile to attempt to retain industrial civilisation, which he regards as doomed, with a new power supply. We think Monbiot was correct to reply that a rapid and unplanned energy descent, with no substitute energy supplies at all, would probably be catastrophic and would carry its own adverse ecological impacts. For example there would be accelerated deforestation as people meet their heating needs. Infrastructure for waste management would deteriorate, enhancing pollution. And food production would be curtailed through fertilizer scarcity.

The energy challenge can be partly visualised via consideration of the relationship between the level of income in a society and the proportion of the population working in agriculture. There is a very tight inverse relationship; virtually no economy with high per capita GDP has more than a small minority of its population in that sector (10% according to Giampietro, 1997). Release of the population from agriculture, enabled by an abundance of cheap fossil energy, appears to have been an essential feature of industrial development. Agrarian economies prior to the fossil fuel era had far less energy surplus after providing for basic needs, and population growth was therefore held in check. If this pattern is suddenly thrown into reverse then a high proportion of the population have, apparently, to return to the land. With an agrarian population the economy would not be able to reproduce its non-agricultural sectors.

The idea that greater efficiency automatically leads to reduced energy use is contradicted by a phenomenon variously referred to as 'rebound,' 'backfire,' the 'Jevons paradox,' and the 'Khazzoom-Brookes postulate.' The problem that all these terms refer to is that when an energy efficiency measure is implemented, the service that the energy consumption provides effectively falls in price. This normally encourages us to use more of that service, but it also makes us better off, enabling us to produce and consume more goods and services generally.

Similarly, the more efficient an industrial process becomes, the cheaper. So the more applications it will find, and to the extent that it brings about savings, these may finance other expenditures and investments. The boundaries of this process are unclear, reaching into the financial system and beyond. Thus, even if a person or business decides to save money and energy, their bank may issue new loans backed by their savings which trigger increased economic activity and energy use elsewhere in the economy, with further knock-on effects. The rebound effect is the total increase in energy use associated with an energy efficiency measure, relative to the reduced level of energy use that would occur with no change in behaviour anywhere. Backfire, or the Jevons Paradox, is when energy use increases overall – that is, when rebound exceeds 100%.

In the absence of a carbon cap, or other brake on energy usage, there appears to be no theoretical reason to expect the overall rebound effect from energy efficiency to be less than 100%. So energy efficiency measures may increase fossil fuel use. Sorrell (2007) estimates the normal extent of rebound to be far lower, at around 30%, but is summarising a literature in which most empirical studies examine only the 'direct' rebound effect. The direct effect is that on the use of the technology which is improved, for example, extra miles driven as a result of a more efficient car. There are few credible attempts at empirical quantification of the total effect. Polimeni *et al.* (2009, ch4) present macroeconomic evidence. The relationship between total primary energy consumption and energy intensity (energy used per unit GDP) is explored, controlling for other relevant factors including population size and density, and the level of GDP. The USA, 16 European countries, Asia, and Brazil are considered. In each case a very strong positive relationship is reported, indicating that greater efficiency is associated with increased use of fossil fuels.

Although as countries industrialise the carbon intensity of each unit of GDP may decrease, this appears, therefore, to be more than compensated by expanded production and consumption possibilities. The (normally) elastic and expansive nature of the financial system, along with the competitive nature of business, it seems, both drive and facilitate the progression of the economy towards the take-up of these opportunities.

It is conceivable that this effect also extends to renewable energy technologies. For example, advocates of wind energy typically claim that a wind turbine produces a quantity of energy over its lifetime scores of times greater than the energy it takes to create it. If the turbines are produced using fossil fuel energy, however, as at present, this may have the effect of increasing the energy efficiency of those fuels, compared to their use in conventional fossil fuel power stations. The result could, in principle, be backfire in fossil fuel use. We do not claim that there is evidence of this, but the general logic of the rebound / backfire effect suggests that it is a disturbing possibility.

With a carbon cap in place, however, the picture changes. Energy efficiency measures could not lead to backfire in fossil fuel use, since a physical limit would be set on the amount of coal, oil and gas that could enter the economy. In addition, by increasing the scarcity of fossil fuels, the cap will increase their prices. This will further abate rebound. Cap and Share would therefore lock in the gains from efficiency measures. The cap would be driving reduced throughput with or without improved efficiency. Efficiency measures make this reduction more tolerable by mitigating hardships that would otherwise result.

To some extent higher energy prices will drive both efficiency improvements and renewables investments. However, there is no guarantee that the price incentives will be strong enough on their own to bring about the required pace of change. A key soure of market failure in both cases is short-termism, underpinned by the interest-bearing debt-based money system. Returns on investment are routinely compared in 'net present value' calculations to the interest that could otherwise be earned on the funds invested, as a default comparison. The higher the rate of interest, the shorter is the effective timescale over which productive investments will be evaluated. Other general sources of market failure include incomplete information and bounded rationality. It may take considerable research to arrive at an informed view on which products or technologies are best. People may also be unaware, or in denial, of the likely future trajectory of energy prices.

If efficiency and energy supply change too slowly, political resistance to the cap may follow as it is driven down, since living standards will fall. Extra measures are therefore called for, both to hasten the pace of change generally and correct for specific problems with free markets. The top level framework rationing fossil fuels and specific technological measures are both necessary, but each requires the other to have the desired effects.

Energy Efficiency Measures

There appears to be significant potential even in developed economies for energy efficiency improvements. Commonly-used 'first law' measures of efficiency mask the extent of this potential. These do not take into account the maximum amount of work that could in theory be done with the resources used. 'Second law' measures do take this into account and are generally much lower. Ayres (1998, tables 2 and 3) for example, estimates that spatial heating in a developed country context by the early 1990s was 72% efficient in the first law sense that 72% of the energy content of the fuel ended up as heat in rooms. Applying the second law measure, the same work could ideally have been done with an estimated 8% of the energy inputs actually used. One can visualise such waste considering an open gas fire, from the fact that one is not generally using it to cook in addition to heating the room, which one can easily do by toasting food on a fork. The waste is incurred in central heating systems by using gas that can burn at temperatures of over 1000°C to heat radiators to 50-80°C; the high temperature flame could be much better utilised. It could in principle power heat pumps, for instance, resulting in less gas use per degree day of heating. For each category of useful work reported by Ayres, second law efficiency is estimated to be a fraction of first law efficiency.

Salient energy efficiency measures include thermal upgrades to buildings and greater efficiency of electric and electronic appliances and vehicles. In addition to raising energy prices, C&S could fund efficiency improvements, particularly to buildings, through the redistribution of revenues to citizens. There are significant problems with free market provision of efficiency improvements though. Energy use may be a relatively small component of costs. Tickell (2008) gives the example of cars. Higher fuel prices give a weak signal here, because for a new car fuel consumption costs are dwarfed by depreciation, the rapid fall in the value of the car over the first few years of use. Buyers of second hand vehicles can only choose from models produced in previous years.

Peculiarities of housing markets cause further inertia. The key problem is 'split incentives'. In rental accommodation, that is, the landlord has only the weakest of incentives to invest to lower the tenant's heating bill. A tenant has no incentive to meet the investment costs, since tenancy is typically a shorter term arrangement than the payback time for the upgrade. These problems are compounded by risk of damage to the building associated with upgrades, the burden of which falls on the landlord. Home-owners face a similar incentive barrier, given that it is probable that they will relocate within the payback period of the upgrade.

This means that in the absence of additional measures, renters would suffer greater exposure to heating cost increases than owner-occupiers, exacerbating already inequitable outcomes in housing. Possible additional measures include legal requirements on landlords to conduct thermal upgrades, coupled with comprehensive upgrade grants, and innovative billing arrangements. Regarding the last option, occupants, whether owners or renters, could pay back the upfront costs of refurbishment by paying a premium on (reduced) energy bills, if the bill remains registered to the property rather than the occupant. That is, on selling up or moving out, the premium would fall on the new occupant.

Energy efficiency standards can be set on goods, appliances and vehicles. These should take the form of minimum efficiency standards, rather than the currently popular but weaker measures such as labelling requirements. Cases can be cited where such schemes have backfired. For example, Tickell (2008) discusses the case of vehicle efficiency legislation in the U.S., which because of industry lobbying did not cover utility vehicles. This allowed manufacturers to produce and heavily market the now notorious Sports Utility Vehicles / 4x4s. We consider such cases to show the need for better designed regulation rather than a case for non-intervention.

Renewable Energy Technologies

For an overview of scenario studies of the potential of renewables globally see Boyle (2007). Most of the studies covered are upbeat about potential. Paticularly optimistic assessments have been released for offshore wind for the UK (PIRC, 2010) and concentrating solar power for Africa (DLR, 2006, which analyses this from the perspective of European demand), but the availability of renewable energy resources is very variable across countries. It is also necessary to take assessments chasing financial backing with scepticism. For more critical reflections on renewables see Smil (2011) and Mackay (2009). Key challenges include the low power density of renewables compared to that of power generation by fossil fuels. That is, the output measured in Watts per square metre is far lower. This is why renewables tend to have exacting land use requirements. There is also a general problem of energy storage to match the time profiles of supply and demand.

If an upstream cap were in place, the relative increase in cost of fossil fuels could render many current incentive schemes obsolete. In the UK, these comprise the complicated existing arrangements surrounding 'Renewables Obligation Certificates,' the Climate Change Levvy and a host of other ad-hoc measures which effectively price CO_2 (Tickell 2008). To this list must be added the feed-in-tariffs currently used to boost investment in solar photovoltaics. The tariffs offer financial incentives for households to invest in PV panels by paying a fee for each kilowatt hour of electricity they generate. This is financed by power generation companies and therefore by all electricity users through their bills, so has attracted criticism because of its regressive effects. Poorer households, that is, are unable to purchase the panels but have the tariff passed on in their bills.

Specific reasons for government intervention include the public good properties of technological innovation, which means that there is a conflict between its rapid adoption and the incentives for the innovator to undertake research and development. This implies a continued role for public funding, and regulation involving time-delimited patents systems. Similarly, some renewables systems require large infrastructural support. Offshore wind, for example, requires extensive capital expenditures to enable the power supply to be connected to a grid. Since such equipment benefits all generators, there is insufficient incentive for individual providers to fund it. In addition, with renewable energy technologies there are fundamental uncertainties about which will ultimately prove viable, and therefore about whether investments in research and development will yield a return.

Whilst many of the reasons for government support of renewables may be familiar, the case is distinctive because there is limited opportunity to get things wrong, lest the remaining fossil fuel budget be expended in support of the wrong kind of technology. Thus, with regard to both subsidies and direct regulation, the saga of biofuels is sobering. Both the EU and US have issued legislation supporting biofuels. In the EU, Directive 2009/28/EC requires that renewables make up a minimum of 10% of transport fuels by 2020, which is mandatory for all EU member states.[10] In the US, the Energy Independence and Security Act 2007 ordered the use of 136 billion litres of biofuel per year by 2022.

Giampietro and Mayumi (2009) examine the feasibility of biofuels on several dimensions. The analysis accounts for the land use requirements of ethanol,

[10] The earlier directive 2003/30/EC explicitly specified biofuels as the fuel from renewable sources in a target of 5.75% of transport fuel. The current target has been received with outcry from civil society groups (Giampietro and Mayumi, 2009, p6-7). It is not clear at present what could substitute for biofuels in meeting it.

based on the per hectare performance of energy crops in capturing solar energy, conversion losses turning crops to fuel, the scale of power requirements to the rest of society, and the internal consumption of fuel by the energy sector itself, within an integrated energetic model of the entire economy. The authors estimate that for even 10% of US transport fuel to be supplied by bioethanol, independently of fossil fuels, 35 times the US arable land currently in production would be required. Similar conclusions hold for the European Union. For example, the authors estimate that for Italy to supply 30% of its transport fuel by biofuels, without fossil fuel inputs, would require 94% of the labour supply to work in agriculture and around 7 times the agricultural land in production.

If this analysis holds, and more summary calculations by Mackay across the range of biofuels (Mackay 2009, p283-286) are supportive, it is scandalous that the biofuels industry has managed to secure a high level of political support, against the background of the food riots of 2007-8 and ongoing global food crisis. The moral of the story would appear to be that there is massive scope for market and government failure in emerging energy markets, exacerbated by the increasingly close relationship between academic scientists and industry.

Measures that could be enacted to prevent the recurrence of such a situation might include the activity of governments in the role of purchasers, and possibly producers, of renewable energy technologies, based on their all-round performance. That is, governments should not limit themselves to enabling market activity. Research governance policies in universities might also be reformed. At present, there appears to be little or no ethical filter on research applications outside of the medical and social sciences for example. This situation might be remedied using mandatory, multidisciplinary research ethics panels for technological research. This might also require reversing the current trend towards increasingly competitive public funding arrangements between academic institutions, if such panels were to have more than face value. The wisdom of allocating substantial proportions of public research funds by matching industry funding should also be reconsidered, given the danger of 'lock in' to inferior technologies.

6. Transport Systems and Relocalisation

Changes in transport systems, and profound changes following from this, seem inevitable, and the need for change would be intensified under a carbon cap. Real distances, measured in the time it takes to get from A to B, have fallen dramatically over time, because of changing transport technologies. Figure 4 visualises this for the case of overseas travel. Traditional, renewables-powered modes were displaced in the nineteenth century by fossil-fuel based vehicles of increasing sophistication, efficiency and power, culminating in jet planes in the 1960s. As a result, intercontinental travel can now take less time than domestic travel before the industrial revolution.

The associated cost reductions, combined with the imperative on producers to expand for reasons stemming from the financial sector, explain much of what is otherwise puzzling about current spatial patterns of production and distribution. They help to explain why even identical produce is exchanged between countries, and why food is shipped back and forth with apparent wantonness before it is sold. Exporting and importing goods, or sending produce up and down the motorway to packaging and distribution centres, for example, might be not much different from cost and time perspectives than producing and selling locally in the pre-1840 world. The diagram presumably understates the true scale of shrinkage between the first three globes, because various other factors impeded long distance travel, including localised political institutions, lack of geographical knowledge and the dangers of long sea voyages.

Cap and Share impacts on transport mainly by hastening the reduced availability of current liquid transport fuels. In the absence of miraculous technological advances, this means driving and in particular flying less, greater use of road and rail, cycling and walking more, plus more localised production and distribution generally. However, as we noted in the context of energy policy, if this came about all at once, without measures to enable change, a high degree of popular, and organised, resistance and unrest can be anticipated. In effect, reduced fossil fuel availability would put global shrinkage into reverse, fragmenting the 'global village' into a scatter of disconnected and over-specialised settlements, progressively less able to provide for themselves.

Moving 'backwards' from the rightmost to the leftmost globes in figure 4 involves increased real distances of a factor of around 65. Given that the sudden and unanticipated arrival of such enhanced real distances is likely to be calamitous, measures are needed both to rationalise transport and to kick-start relocalisation pre-emptively.

Included could be measures such as Alan Storkey's proposal for a rationalised coach system, with freight going onto rail (Storkey 2007; Monbiot 2007; CAT 2010). Although developed for the UK the principles are general. Coaches reduce congestion by displacing cars, and reduce fuel consumption because of elimination of duplication and economies of scale. Storkey estimates that each coach with normal occupancy represents one mile of car traffic travelling at 60mph. The core proposal is for a conjunction of orbital coach services, linking major population centres, and bus routes connecting the inter-city network with towns, city centres and suburbs. Use of such services increases with the frequency of provision. Such a rationalised network would achieve both direct and indirect fuel savings, the latter through reduced congestion.

Further savings might be made in combination with technological measures, such as electric coach fleets, if the efficiency claims of their proponents can be justified on a life cycle basis (CAT 2010). The main uncertainties surrounding

Figure 4.

Period	1500-1840	1850-1930	1950s	1960s
Technology	Sailing ships	Steam ships	Propeller planes	Jet planes
Fastest Speed (mph)	10	36	300-400	600-700

Shrinkage of the globe in terms of 'real distance,' because of increasingly powerful fossil fuel transport technologies. Distances, such as the circumference of the earth, are proportional to the shortest possible overseas travel times, determined by the fastest speeds available shown in mph. Adapted from Dicken (1998), figure 5.3.

such claims stem from scarcity of such analyses, and potential depletion of specific resources that their widespread adoption would exacerbate – in this case, supplies of heavy metals for batteries.

In support of the coach network proposal, we note that Cuba in the 'special period' heavily increased use of coaches, including improvised tractor trailers ('camels') adapted to hold large numbers of passengers. Coach travel is able to expand significantly and to substitute rapidly for cars, without substantial infrastructure expansion, in contrast to rail. However, there is limited gain from modal shift (cars to buses and freight to rail) alone. Given the very low levels to which emissions have to fall what is ultimately needed is less reliance on motorised transport per se (Mobbs 2004).

Pre-emptive relocalisation can to some extent be furthered by action to curb externalities such as waste, in particular from product packaging and disposal. Heather Rogers (2005, p134-137) details how the bottled drinks industry is able to extend its spatial scale through externalising costs of waste disposal onto taxpayers. Where regulations exist forcing companies to use refillable containers, this acts as a constraint on the ability of the manufacturers to centralise and extend their operations. Delivery lorries have to return with the empty containers, for example. Sales expansion therefore requires replication of production and bottling facilities. This results in a pattern of localised production and consumption, in addition to achieving efficiency gains in terms of energy and raw materials. Unfortunately business has often lobbied effectively to remove such regulations, since they impact on profits and growth in market share. This has enabled increasing concentration of the drinks market in the hands of large corporations.

In some countries, including the UK, there has been a revival of interest in local production and sale of foodstuffs. This has been triggered to some extent by producers' being squeezed by the market power of large retailers, and supported by grassroots movements such as the Transition Towns. However, these have met with limited support from authorities. Measures authorities could take that would bolster local activity include planning restrictions on maximum size of retail outlets, rent controls on high street stores, and a supportive stance towards new food markets. For example, in France local producers are often allowed to sell in hypermarket car parks.

7. Land Value Tax

LVT is taxation levied on the value of land, raised either through a regular (say yearly) levy on its rental value, or as a percentage of its sale value. LVT has strong economic arguments in its favour, set out in the works of Henry George – see for example George (1879). It is a non-distortionary tax, that is, one which does not interfere with supply, since the supply of land is effectively given. Since land's market value is social in origin, arising from the competing uses to which people might put it, rather than the result of useful activity of the landowner, taxing it seems just. Moreover, since access to land is a necessity for living, all people who do not own land are subject to rent payments, in the sense of payments of unearned income to landowners. It follows that there is significant potential to use land values as a key, if not the, basis of taxation. George argued that LVT should replace all other taxes, including income tax, and some modern proponents including Fred Harrison (2008) also argue in this vein. Others hold that LVT is insufficient and that other sources of rents (in the sense of unearned income) exist and should be tapped to fund the public purse. Cap and Share is, indeed, an example of this. For a recent report on land value taxation, covering the practicalities of implementation in an Irish context, see Smart Tax Network (2010).

LVT fits with the measures outlined above as follows. Firstly, if incentive payments to landowners or farmers are used as part of a nation's Carbon Maintenance Fee strategy, market values for land with carbon storage and sequestration potential will increase, even with no actual improvements made. LVT would counteract increased land values being realised as windfall gains, which is inequitable, and would provide a robust tax base to help finance the CMF. Secondly, in LICs, where a global C&S would result in large transfers from high consumption countries, it is likely that people will wish to buy land with their permit revenues. This will result in windfall gains to existing landowners if the increase in values is not captured for the common good.

Potential concerns with LVT include weakening CMF incentives to maintain and improve carbon sinks, and possibly increasing the development of green space. Consider the first issue. If, as we argued in section 1, downstream payments to farmers or landowners would realistically be based on activity monitoring, land values will rise to reflect any potential subsidy earnings. This is independent of whether the subsidy is in fact earned, so the incentives for carbon maintenance and sequestering activities are unaffected.

Alternatively, there might be payments based on the carbon content of specific landholdings, if measurement technologies improve. It is common for advocates of LVT to recommend taxing the unimproved value of the land, so as not to discourage improvements to buildings (Daly and Cobb 1994, p235). Transferring this proposal to the present context, 'unimproved' should also be understood in carbon terms. Lands will increase in value to an extent that could in principle be estimated even if no carbon improvements are actually made, because of the potential to use the land to gain CMF-related payments. This leaves the incentive to earn the subsidy intact.

For leaseholders, the rental value of the land is not affected by LVT, merely who ends up with the rents. Therefore any CMF-related incentives accruing to them would be undiluted. Competition between renters could be expected to bid away the unearned portion of these payments in higher rents, just as it would in the absence of LVT.

Regarding the second issue, LVT may have mixed effects on green space. One of these is to ease pressure on any land reserved as greenbelt, through making urban space more readily available, since it becomes costly to hold empty land and property. A second potential effect, which pushes in the other direction, could be to jeopardise wilderness and therefore biodiversity, since all land holdings will also have to pay enough to cover any tax owing. Much currently unused land will have low market value, however, and therefore attract lower rates of LVT. Also, we have argued that, for wilderness land with carbon value, the CMF scheme will still operate as an incentive in the presence of LVT. It is possible for there to be conflict between use of land for carbon sequestration and biodiversity, however. For green space in or near cities, the CMF value of the land is likely to be dominated by property development value. The government will in this case share an incentive for the land to be developed, since the value of land, and so tax, with planning permission greatly exceeds that without it. It is likely that there are other cases where this problem would arise. For example, for unused land with both strong agricultural potential and high ecological value the former may dominate.

These considerations underline the fact that other elements of the policy mix, such as spatial planning controls need to be used to counteract any undesirable side-effects of the policy. The tax is also flexible; a zero rate could apply to land below a certain value for example, to assist farmers on low incomes.

8. Less Industrialised Countries

The discussion above focuses primarily on industrialised countries' policies, as opposed to those of LICs. This is partly a reflection of our focus of climate change mitigation, which is primarily the industrialised world's responsibility. Policy discussions often focus on adaptation in an LIC context. However, if only industrialised countries adopt C&S, the fossil fuel based economy may quickly migrate to LICs. This has already been happening in India, China and other eastern countries to such an extent that in many respects labels such as LIC have become questionable. LICs therefore need to be part of the constraining framework. Most of the populace of such countries stand to benefit from C&S, for perhaps most of the duration of the scheme, since people in the South typically consume so little per capita compared to their counterparts in the North. Also, much of the potential for enhancing carbon cycles and sinks lies in LICs.

Adverse effects of climate change are likely to be born disproportionately by LICs; some small island states, such as Tuvalu in the South Pacific, are already becoming uninhabitable because of sea level rise and resulting salination of agricultural land (Lynas 2004). The only available response in those circumstances would appear to be managed migration, with the cooperation of neighbouring states. Other reasons why many LICs may suffer disproportionately from climate change include narrower tolerance of crops to increased temperatures, outside temperate zones. Lower per capita income reduces resilience to loss of crops because it implies less access to imported food.

LICs could receive an enormous boost from a global cap and share scheme (Wakeford 2008), helping with such costs. However, this could either be via per capita shares from individual sales of permits or a managed fund (as under Cap and Dividend proposals), with possibly distinct implications for how the resources would ultimately be used. These options are discussed in chapters 6 and 7. Having said this, the third world debt crisis continues to be a major drain of resources for the South. Further annulment of 3rd world debt therefore seems essential lest transfers of Cap and Share scarcity rents be diverted to interest payments transferred back to the North. In the same vein, LICs will benefit more from such revenues to the extent that they avoid an economic trajectory determined by imported fossil fuels. The 1970s oil shocks, it should be noted, played a key role in fomenting the debt crisis. This resulted in foreign exchange shortages in LICs and petrodollar gluts in OPEC countries, stimulating lending by the latter to the former.

We believe much of the discussion in previous sections, therefore, is in fact no less relevant to the LICs. In many LICs agriculture has become petroleum dependent, so tooling up, and skilling up, for organic agriculture and other agro-ecological initiatives seems a policy priority. Debt-based money has the same dynamic in LICs as elsewhere, in conflict with hard energy constraints, and so debt-free alternatives need to be developed in this context too. In addition, one likely use of C&S funds by individuals, or communities under cap and dividend, is to buy land. Land values could be predicted to increase, giving rise to windfall gains for landowners and speculative bubbles. LVT would help by capturing speculative gains and enabling the redistributing of windfalls. A barrier to certain policies at present, including Cap and Share and LVT, is low administrative capacity compared to the North, so strengthening this would be a priority.

The geopolitical situation of LICs seems to be a significant barrier to enlightened policy, however. Were they to tackle the issue of the lease or sale of agricultural lands to foreign companies, for example, they would face conflict with the IMF, World Bank and WTO, whose neoliberal ideology favours the land deals (GRAIN 2008, 2009). These institutions have operated in the interests of transnational corporations and their political backers, undermining national economic sovereignty. Often, for example, they have lent large sums to countries in financial difficulty with onerous conditions that both precluded their repayment and favoured western corporations and strategic interests. Tariff barriers were reduced, domestic industries privatised and sold, cheap access to raw materials was provided to overseas firms and so on. This caused the countries in question to take on further loans, deepening their problems and debt dependency which can be exploited by western government and business interests (Korten 1996). To take a stand against these institutions calls for considerable bravado, as amongst other sanctions they can make it impossible for a nation to borrow, and crash its currency.[11] There might also be diplomatic problems and intimidatory tactics or worse against the politicians involved, as recounted in Perkins (2005) for the case of Latin America.

[11] Far from resisting the transfer of land rights to transnational corporations, governments pursuing development are frequently assisting this process. See Shiva (2011) on India, for example.

9. Conclusion

A CO_2 policy package based around Cap and Share, we believe, should include monetary reform, the CMF, state support for renewable energy and energy efficiency measures, land use controls and restrictions on biofuels, an agricultural support program promoting organic, low input and agro-ecological methods, and rationalised transport infrastructure. Land value taxation would dovetail with this set of policies and provide a durable funding stream for the CMF. It also captures rents for the public purse in what are likely to be difficult times economically.

The policy set presented has, we believe, an encouraging internal coherence, but nonetheless we finish on a cautionary note. This is because we have not considered all aspects of the problem under consideration, partly because of space and time constraints and partly because of limiting ourselves to consideration of the CO_2 aspect of the climate change problem, and by extension the climate change dimensions of the ecological problem. We may therefore fall victim to the so-called "fallacy of misplaced concreteness" which results from treating an abstraction as if it were the phenomenon of interest in all its fullness.

Not addressed here include, for example, questions of reform to organisational forms and corporate law. In particular there is the issue of the 'fiduciary obligation' of company directors to shareholders, commonly interpreted to override any social and environmental considerations where these conflict with profitability.[12] We have also abstracted from the need for action in relation to the other greenhouse gases, including Nitrous Oxide, Methane, and the Halocarbons, and the effects of black Carbon particles. Finally, we have also abstracted from other aspects of the ecological crisis, including the need to preserve biodiversity, and the question of human population growth.

---

12 See for example Milton Friedman (1970) for a classic statement of this interpretation.

References

1. Ayres, Robert. 1998. Technological Progress: a Proposed Measure. Technological Forecasting and Social Change 59, 213–233.
2. Azeez, Gundula. 2009. Soil Carbon and Organic Farming. Soil Association report.
3. Byrne, Corinna. 2010. Refocusing the Purpose of the Land: from Emissions Source to Carbon Sink. *Fleeing Vesuvius*. Dublin: FEASTA
4. Bartlett, Albert A. 1978. Forgotten Fundamentals of the Energy Crisis. *American Journal of Physics*, 46, 876-888.
5. Bowles Samuel, 2008. Policies Designed for Self-Interested Citizens may Undermine "the Moral Sentiments". Science 320: 1605-1609.
6. Boyle, G. 2007. Long Term, Renewables-Intensive World Energy Scenarios. In D. Elliott (ed.) *Sustainable Energy, Opportunties and Limitations*. Palgrave Macmillan.
7. Bruges, James 2009. The Biochar Debate. Green Books: Totnes.
8. CAT 2010. *ZeroCarbonBritain* 2030. Centre for Alternative Technology.
9. CCSN 2010. *Reducing Greenhouse Emissions from Activities on the Land*. Carbon Cycles and Sinks Network, working paper.
10. Daly, Herman E. and Cobb, John B. 1994. *For the Common Good*. 2nd edition. Boston: Beacon Press.
11. Dicken, Peter. 1998. *Global Shift*. Third Edition. Paul Chapman Publishing.
12. DLR (2006). Trans-Mediterranean Interconnection for Concentrating Solar Power. German Aerospace Center (DLR).
13. Douthwaite, Richard. 2010. The Supply of Money in an Energy Scarce World. In *Flight from Vesuvius*. Dublin: FEASTA
14. Friedman, Milton. 1970. The Social Responsibility of Business is to Increase its Profits. *The New York Times Magazine*, September 13.
15. George, Henry. 1879. *Poverty and Progress*. New York: Doubleday, Page and Co.
16. Giampietro, Mario. 1997. Socioeconomic Pressure, Demographic Pressure, Environmental Loading and Technological Changes in Agriculture. *Agriculture, Ecosystems and Environment* 65, 201-229.
17. Giampietro, Mario and Kozo Mayumi. 2009. *The Biofuel Delusion*. Earthscan.
18. Gneezy, Uri. and Rustichini, Aldo 2000. A Fine is a Price. *Journal of Legal Studies*, 29, 1-17.
19. GRAIN 2008. The 2008 Land Grab for Food and Financial Security. GRAIN briefing, October 2008. http://www.grain.org/briefings_files/landgrab-2008-en.pdf
20. GRAIN 2009. Rice Land Grabs Undermine Food Security in Africa. Against the Grain, Jan 2009. http://www.grain.org/articles/?id=46
21. Gudmundssen, I. 2010. Address to the Bank of International Settlements.
22. Harrison, Fred. 2008. *The Silver Bullet*. The International Union for Land Value Taxation: London.
23. IAASTD 2008. Agriculture at a Crossroads.
24. Korten, David. 1996. *When Corporations Rule the World*. Kumarian Press: Connecticut.
25. Lynas, Mark. 2004. *High Tide*. Macmillan.
26. MacKay, David. 2009. *Sustainable Energy without the Hot Air*. Cambridge: UIT Press.
27. Minsky, Hyman 1982. *Can 'it' Happen Again? Essays on Instability and Finance*. M.E. Sharpe: Armonk, N.Y.
28. Mitchell, David. 2008. A note on rising food prices. Draft report prepared for the World Bank
29. Mobbs, Paul. 2004. *Energy Beyond Oil*. Trowbridge: Troubador.
30. Monbiot, George. 2007. *Heat. How can we Stop the Planet Burning*. London: Penguin.
31. Perkins, John. 2005. *Confessions of an Economic Hit Man*. London: Ebury Press.

32. PIRC 2010. The Offshore Valuation. Public Interest Research Centre Report on behalf of the Offshore Valuation Group.
33. Polimeni, John M., Kozo Mayumi, Mario Giampietro and Blake Alcott. 2009. *The Myth of Resource Efficiency.* London: Earthscan.
34. Rogers, Helen. 2005 *Gone Tomorrow: the Hidden Life of Garbage.* New York: The New Press.
35. Rowbotham, Michael. 1998. *The Grip of Death.* Charlbury: John Carpenter Press.
36. Shiva, Vandana. 2011.The Great Land Grab: India's War on Farmers. *Al Jazeera*, June 7th 2011
37. Smil, Vaclav. 2011. Global Energy: The Latest Infatuations. *American Scientist.* 99, 212-219.
38. Sorrell, Steven. 2007. The Rebound Effect. UK Energy Research Centre report. Available at http://www.ukerc.ac.uk
39. Stern, Nicholas. 2006. *The Economics of Climate Change: the Stern Review.* Cambridge University Press: Cambridge.
40. STN 2010. *Implementation of Site Value Tax in Ireland.* Smart Taxes Network
41. Storkey, Alan. (2007). A Motorway Based National Coach System. Bro Emlyn – for Peace and Justice.
42. Tickell, Oliver. 2008. *Kyoto2.* London: Zed Books.
43. Titmuss, Richard. 1970. *The Gift Relationship.* London: Allen and Unwin.
44. Wakeford, Jeremy. 2008. *Potential Impacts of a Cap and Share Scheme on South Africa.* Foundation for the Economics of Sustainability.
45. Werner, Richard. 2004. *New Paradigm in Macroeconomics.*
46. Wright, Julia. 2009. *Sustainble Agriculture and Food Security in and Era of Oil Scarcity.* Lessons from Cuba. London: Earthscan.

Chapter 5

Operating effectively at the world level

John Jopling

"The whole idea of operating effectively at the world level still seems in some way peculiar and unlikely. The Planet is not yet the centre of rational loyalty for all human kind". 1972 Report on the Human Environment for the first UN Conference on the Human Environment

Introduction

Given the number and nature of the global problems facing humanity today, not least climate change, I believe that human kind's most crucial need now is to have the capacity to "operate effectively at the world level". This therefore is the subject of this chapter. My starting point is that we do not at present have this capacity. I want to suggest how we might acquire it. Like Richard Douthwaite in relation to economic issues, I think there are grounds for optimism here too: the idea of being able to do this is not now so "peculiar and unlikely" as was considered in 1972. This is because I think there are now indications that the Planet is now becoming "the centre of rational loyalty for all human kind". I hope we can now learn to operate effectively at the world level.

This chapter starts by explaining why the current system of global governance and the process for addressing climate change in particular – that of negotiation between the governments of nation states – has failed and indeed was bound to fail: I point to some of the systemic elements in the design of the system which prevent it from being able to operate effectively. But the problem is much wider than the particular process being used to address climate change. This section ends with some observations about the currently dominant concept of governance more generally: a new climate regime will need to manifest a completely different paradigm of governance.

I then suggest some of the characteristics of a new system of effective global governance. I contrast the advantages of the suggested new format with the shortcomings of the existing one. I go on to explain why I believe that creating a new system is not only necessary but also possible; and that it lies within the powers of ordinary people, who are effectively excluded from the current dominant system and cannot hope to influence it, to step-by-step build a new global structure, making sure from the outset, and at each stage of its development, that it is designed to be capable of operating effectively as a global system.

Our present powerless state is due to trying to operate within a defective system. This proposal aims to address that problem, by gradually building a suitably designed system.

The incompetence of the UN Framework Convention on Climate Change (UNFCCC) and of today's governments generally

The word 'incompetence' is used in the sense of meaning in effect 'simply not being capable of doing what is needed of it'. The term applies not only to the intergovernmental negotiation process for addressing climate change but to the existing model of government itself: today's governments are systemically incapable of coping with any of the major problems facing humanity today.

An incompetent process

In relation to climate change, we can take as given the fact that the current governmental and intergovernmental systems for addressing this extraordinary problem have failed and show no signs of recovering. This is now widely acknowledged. 25 years after the problem of the rising level of global warming gases in the atmosphere was brought to the attention of governments, the international community has failed even to agree what is a safe level, let alone commit to achieving such a level. The 1992 UN Framework Convention on Climate Change was basically a framework convention that left much, far too much, to be negotiated later. Whilst the stated objective was to stabilise greenhouse gas concentrations, no target level was stated, nor has one ever been agreed. Since 1992 net emissions have continued to increase, at an increasing rate too, resulting in continuing temperature and sea level rises and triggering various dangerous positive feedback loops, such as the melting of arctic ice and the warming of frozen ground in Siberia. Climate scientists are in no doubt that we now have a crisis [1].

I attribute the failure of the UNFCCC not to any incompetence on the part of the individuals involved in operating the system but to the model of intergovernmental politics which they inherited and on which the current systems are based. The problem is not the behaviour of particular nations or the people representing them (which is where the blame is usually leveled), but the model. The model has many flaws:

World as collection of states

It is based on the view of the world as a collection of states. This means that international action on climate change depends on agreement being reached through negotiations between the governments of countries with widely differing circumstances and widely differing, and often conflicting, interests in the context of climate change. The obvious and now widely recognised result is that it is extremely difficult, perhaps even impossible, for the nations of the world to agree about something as contentious and complicated as climate change and what to do about it [2].

Agreement only by compromise

Agreement, if it is reached, can only be achieved by compromise. So the aim of those in charge of the process is to secure a compromise, rather than for governments to take the actions needed to avoid the disaster we are currently heading for [3].

Limited powers of governments to deliver

Another simple but frequently overlooked point is that governments have limited powers: nation states are limited in the degree to which they can directly affect emissions of greenhouse gases and the ability of societies and economies to adapt to climate change [4]. Even if they agree something between themselves, it does not necessarily follow that it will happen. Which incidentally tends to mean that they will only agree what they think they can achieve without difficulty.

Other species and ecosystems not parties

A fundamental flaw is that there is no adequate representation of other species and ecosystems or of future generations of our own species. If any compromise is arrived at, it will necessarily have been agreed without either future generations or other species being represented round the table. These interests are of course represented by numerous NGOs given a hearing in the negotiations but these organisations are not going to be parties to any agreements reached and their representations are in practice generally ineffective.

Governments able to interfere in the science

Another problem is the system established for giving governments and the public information about climate change. The International Panel on Climate Change (IPPC), the system set up by the United Nations Environmental Programme (UNEP) and the World Meteorological Organisation (WMO) in 1988, is neither comprehensive nor independent: governments can and do influence the terms of the published reports [5].This creates the perfect environment for fossil-fuel funded propaganda to flourish in. The result: the public is misinformed and bewildered.

No risk management

Linked to that is the absence of an effective risk management system to assess the likelihood of particular changes and the seriousness of the consequences. Risk management is essential for applying the precautionary principle, which although enshrined in the UNFCCC is currently being largely ignored due to the lack of such a process within the current arrangements.

No procedure for an emergency

A more specific defect of the UNFCCC process that is directly related to one of the main topics of this book is that it lacks any procedure for taking urgent action in the event of an emergency. Climate change was recognised by the UN General Assembly in 1988 as 'a common concern of mankind' and this was the general understanding at the time the UNFCCC was set up in 1992 and when the Kyoto Protocol was signed in 1997: it was not seen as a crisis requiring immediate action. Since then, due to human-induced forcing, the emission of CO_2 in particular, having triggered a number of positive feed-back systems including the melting of Arctic ice and release of methane from permafrost, the situation has become far more critical. We now have an emergency on our hands: what we need now is urgent action to staunch the haemorrhage of global warming gases into the atmosphere. The UNFCCC was not designed to provide this sort of emergency operation and shows no sign of doing so.

The only possible conclusion is that the current system of negotiation between nation states is not a system that (to repeat the language of the 1972 reporters) enables human kind "to operate effectively at the world level".

I should add that the view that the international system itself is incompetent is neither radical nor new [6].

It must also now be borne in mind that the incompetent international system set up in 1992 is no longer the only actor on the stage of climate change governance. As a number of writers have observed recently, the landscape today is not simply a failing or failed multilateral negotiation process, with the rest of the canvas blank. Many initiatives of many kinds, within the meaning of 'governance' broadly defined, are responding to climate change and to the failure of our governments to address it: local communities, municipal and provincial and national governments, businesses and various other actors are now engaged in a host of projects and processes that are independent from the intergovernmental process and from national regulatory measures [7].

The result is that missing from all existing initiatives in climate change 'governance', whether on their own or together, are a number of critical capacities:

- The ability to set specific global science-based targets, whether in terms of concentrations of particular gases in the atmosphere, or of reducing world emissions into the atmosphere of particular gases, or of increasing particular forms of draw-down, or otherwise. The need for global targets arises simply because the atmosphere is global: climate change is a global phenomenon.
- The capacity to design global policies that are both effective to meet the science based targets and also socially just.
- The capacity to implement and administer policies requiring some form of global administration.

An incompetent model of government

A more fundamental issue concerns the model of government dominant throughout the world today. The failure of governments to reach agreement concerning climate change is not their only failure: they have proved incapable of dealing with a whole range of problems facing them today. In spite of all the advances in science, technology and other fields over the last half century or so, governments, whether democratic or otherwise, and whether or not supported by huge and highly trained civil services, are faced with ever larger and more intractable problems. Government itself seems to be the problem.

This is not the place to analyse the nature of the present governmental system in any depth. Roy Madron and I attempted to do this in our Schumacher Briefing *Gaian Democracies*. We identified 6 core components of the system which together define it:

- first, and most fundamental because it is the purpose of any system that limits the way the system is capable of working, the 'common purpose' of the present dominant governmental system is that of money growth in order to maintain the bank-created, debt-based money system (which is why we named the system a 'Global Monetocracy') [8];
- the system is held together by an 'elite consensus' upholding the values and assumptions of the Monetocracy, which means that alternatives to the growth imperative of the current money system cannot be seriously discussed;

- it justifies itself through a number of shared operational theories including neo-liberal economics, national sovereignty, representative democracy and command-and-control leadership;
- it has a global leadership cadre covering politics, finance, corporate business, academia and the media;
- it operates through the big business – government partnership by means of an armory of operational instruments including transnational corporate capitalism and international institutions, like the World Bank, the IMF, the WTO and the EU [9].

Nor is this the place to describe the wider consequences of having such a system dominate our world, both within nation states and at the international level. Writers such as Noam Chomsky, John Pilger, Colin Leys and Naomi Klein describe precisely the kind of activities and behaviours that are only to be expected from such a system, because they are the inevitable consequences of the design of the system. Given its nature and systemic purpose, it is simply incapable of dealing with circumstances that require humanity to halt growth. There is no way, under the present system, that human activity on the Planet could be managed sustainably, because that would conflict with the growth imperative.

It is hardly surprising that the schemes set up within the UNFCCC have been of little or no use in addressing climate change, but have successfully created several new market opportunities for financial markets, as these help to create money-growth [10].

The crucial point I am making here is that the government systems that have grown up over the last few hundred years are systemically incapable – that is to say incapable because of their design, or more specifically their systemic purpose – of coping with climate change or, for that matter, numerous other 'wicked' problems facing the whole human family at the start of the 21st Century, let alone the further problems that will inevitably confront us in the future due to climate change, oil and gas peak, biodiversity loss, numerous kinds of resource depletion and many others [11]. The systems we have today are holding us back from being able to "operate effectively at the world level". They are actually taking us in the wrong direction and getting in the way.

For many, including many experts in the field of climate change who know that urgent action needs to be, and can be, taken to address climate change and who are appalled by the failure of governments to respond to the evidence placed before them, the problem appears to be a lack of political will on the part of politicians and bureaucrats dominated by armies of corporate lobbyists. This is extremely frustrating because there appears to be no way round it. Public opinion is easily swayed by the media, most of which is also dominated by corporate interests; and anyway it is very difficult for ordinary members of the public, even including distinguished scientists, to have their voices heard on issues such as climate change, certainly not heard above the much louder voices of the corporate lobbyists [12].

The need for a new strategy

If I am right in saying that the human family does not at present have an effective system "for operating effectively at the world level", the question we have to ask ourselves is: what can we, as ordinary citizens of the world, do about it? The current strategy of the most concerned and enlightened people is to increase the pressure on governments. But, given the nature of today's governments, and the dominance of the corporate lobby, I believe that this strategy is extremely unlikely to be effective: it is like pushing a piece of string, attempting to get a system to do something it is incapable of doing.

I conclude that we must develop another strategy, one that has a better chance of success. If, as I believe, the fundamental problem is the system, then to my mind, the only strategy that has any chance of working is that of creating a new system, one that is consciously designed to avoid the flaws in the current one. I suggest that climate change may be the issue that could generate the critical mass of public support needed to build a new system; and that the current window of opportunity to avoid runaway climate change offers an opportunity to make a start on creating a new system that is specifically designed for managing the human response to climate change.

The proposition that new institutional arrangements are needed to manage a global economy operating within global limits is not a new one [13]. I go further and argue that, step by step, we need to build a completely different **kind** of system.

The nature of the proposed new structure

A new structure created by ordinary citizens would have to be constructed from the bottom up and would have to be built brick by brick. There are a number of things we need to get right from the start. We can identify some of these by reference to my list of the systemic defects of the current system. What would be the corresponding features of an effective system?

The world is not just a collection of states, it consists of people and hundreds of different kinds of organisation

My first point was that the UNFCCC and the negotiations conducted under it as well as the Kyoto Treaty assume that the world is a collection of nation states, which therefore have to reach agreement in order for there to be a global decision. Well, the world does not have to be, and should no longer be, thought of simply as a collection of nation states.

In his 1989 Hull University Josephine Onagh Memorial Lecture *International Law and International Revolution: Reconceiving the World,* Professor Philip Allott described how the view of the world as a society of states only arose in the 18th century. Up to then, there was an idea "that all humanity formed a sort of society and that the law governing the whole of humanity reflected that fact". The Spanish writer Francisco de Vitoria, for example, in the 16th Century, regarded this universal law for all human beings as found in natural reason, the rational character of human nature.

That view gradually disappeared as what in fact came into being were nation states ruled over by governments, not world-wide institutions. In the 18th Century, the French writer, Emmerich de Vattel, concluded from this that there could be no great republic set up by nature herself. In his view, the state was not only the internal organisation of the public realm of a society, it was also the whole of a society when seen externally. So states, represented by their governments, were the only legitimate players on the international stage. Vattel was widely read. His thinking suited the interests of the powerful. His was "a book which formed the minds of those who formed international reality, the international reality which is still our own reality today".

The world-of-nation-states point of view still dominates mainstream thinking, though by no means as comprehensively as before, as evidenced by the mushrooming of International Organisations and the number of universal problems such as over-population, starvation, pollution and destruction of the environment confronting mankind [14]. But we do not have to think of the world simply as a collection of nation states. We do not have to assume that measures for addressing global problems, such as climate change, can only arise out of negotiations between nation states.

Climate change directly affects, and is effected by, people – with their cities, their industries, their transport systems etc -, rather than nation states as such. The stakeholders include everyone alive today (especially the vast majority of the world population who are suffering most of the damage caused by climate change but who contributed little of the greenhouse gas emissions), future generations of people and also other species and ecosystems. We need to design a governance system that gives all those affected a voice and is effective to control the activities needing to be controlled.

The basis of climate change targets and policies should be climate science and sound economics, not negotiation and compromise

My next point was that decisions reached by negotiation between nation state governments invariably mean compromise – with targets and commitments that fall short of responsible risk management. The targets and policies should be based on what is necessary for global safety. That will not arise from negotiations but from a dispassionate review of the climate science. The fact that governments may not be prepared to accept those realities is an entirely separate matter – one needs an organisation that is bound to base itself on the climate science as far as possible and that is obliged to tell this truth to all forms of power. If governments do not then accept that, that is an issue to be worked on certainly, a problem that has to be overcome – but that is quite different from relying on negotiated agreements that are inadequate from the point of view of climate safety and the future of humanity. We need a system that is prepared to take a position completely independent of governments and corporations and base itself on the climate science and on its accumulated moral authority, even if that meant, in extremis, an absence of endorsement by a host of governments that are in denial and in dereliction of their duty to their citizens and to world citizens.

The system needs to include all organisations who can deliver

Effective 'command and control' government depends on being able to command and control, if necessary by force. The third point made above was that there are limits to what a modern government can enforce, except perhaps when under threat of invasion. We need a global structure that does not have this problem because it does not depend on having powers of command and control. It will need to operate in a completely different way. This is discussed below.

Other species and ecosystems must be represented

I envisage a new structure that sees itself as representing the whole of humanity including future generations; and, as we depend on, and indeed are part of, the Earth's biodiversity, it will also represent other species and ecosystems. There are many practical ways in which this representation can be realised in the procedures adopted.

The science must be free of government interference

My next point referred to the flaws in the IPCC. To ensure that all decisions are in future based solely on the latest climate science, it will be essential to establish an independent worldwide climate science information service as part of the new structure.

Risk management

Effective risk management is crucial. This should be made mandatory within the new structure.

Taking emergency action

Finally the new structure must be designed so as to be able to reach decisions and act as quickly as the situation requires.

What is the alternative to the Global Monetocracy?

My list of the systemic defects of the current system was followed by a discussion of the current model of government generally. The proposed new structure obviously needs to be based on a completely different model. Any form of 'command and control' system is out of the question. We are looking for a system that is stable but has the flexibility and diversity needed to cope with the complexity of the world and the 'wicked' problems we face.

Here again we can contrast the components of the Global Monetocracy with the kind of system now required to operate effectively at the world level.

- instead of a 'common purpose' resulting in the imperative of economic growth, we need the common purpose of climate safety and climate justice – we need a system that has the Planet "at the centre of rational loyalty for all human kind";
- instead of an 'elite consensus' upholding the values and assumptions of the Monetocracy, we need an atmosphere of tolerance, cooperation and creativity so that every issue that arises can be openly explored and discussed;
- instead of theories like neo-liberal economics and national sovereignty, we need to substitute principles such as subsidiarity and concepts such as human rights, social justice and Earth Jurisprudence (this is referred to below);
- instead of the current global leadership cadre we need a far more open society;
- and instead of the big business – government partnership, we need to exclude corporate influence over the governance system.

These bullet points suggest an outline of the very different system of governance that is required to enable humanity to "operate effectively at the global level". A number of principles of effective organisation developed over the last half century or so may help to flesh it out. Some examples are referred to in the text box.

The principle of subsidiarity

This is the principle that all functions should be carried out at the lowest level at which that function can be carried our satisfactorily [15]. The principle seems right intuitively, giving everyone the maximum freedom compatible with the wellbeing of the wider communities they are part of. This might be the guiding principle for the level at which decisions are taken about an issue discussed elsewhere in this book, the distribution of the proceeds of sale of climate permits. The principle accepts that some functions and decisions may have to be carried out or taken at the global level: putting together the findings of scientists working in a wide range of disciplines relevant to climate is one example; a global cap on the introduction of fossil fuels into the economy and the global level administration of global schemes like Cap and Share are also examples.

Stafford Beer's Viable Systems Model

This model envisages that, instead of responsibility and power being in one entity called the government, autonomy and responsibility are shared out throughout the organisation with the aim of ensuring that it can survive in a changing environment. The VSM offers a language to help people work out how to do this [16].

Ashby's law

This states that only variety can absorb variety. It is one of the laws of the science of self-regulation known as cybernetics, which has been developed as a tool to help organisations manage themselves effectively. For Beer and other cyberneticians this law explains why top-down government, where decisions affecting many are taken by a few, is so ineffective: the few decision-makers do not have the variety to match the variety of the world they are up against, so they are overwhelmed by the complexity of the system [17].

The growing commons movement

Commons thinking, referred to elsewhere in this volume, is much more than a mere theoretical model, it is a living and growing movement which should in future underpin all governance issues relating to natural resources. We need to learn how principles successfully applied to the management of human relationships with local natural resources can be applied to the management of humanity's relationship with the Earth's climate.

Participatory democracy

Whilst representative democracy is the form of democracy that suits command and control government within the Global Monetocracy, the new structure will need to develop forms of participatory democracy. We need to learn, for example, how processes developed to enable thousands of citizens to take part in decision making relating to their city can be developed to enable billions of global citizens to take part in decisions about the use of scarcity rents from the sale of permits to pollute a global resource; and how the practice of courts of law where an advocate is appointed to represent the interests of unborn persons, infants and persons of unsound mind, can be developed to ensure that the interests of those who cannot participate actively are taken fully into account [18].

The principle of participation has another side to the coin: corporate bodies, whose overriding purpose is to make money for the owners, should not be participants, because they are merely legal constructs whose overriding purpose is incompatible with Humanity's interest in remaining a viable species [19]. The distorting consequences of corporations being able to influence governments are evident today almost everywhere, including in the UNFCCC process itself [20].

Reference to the principles referred to in the text box, to the commons movement and participatory democracy is intended to do no more than give a flavour of the sorts of new structures envisaged here. It should be clear that I am not contemplating a 'revolution' in the sense of destroying or undermining any existing governmental or other system. The new framework is envisaged as operating alongside existing systems and working in cooperation with them, so long as they last. The concept needs developing with the help of experts in organisational development [21].

Comparable initiatives and related movements

The idea that an international institution could arise from a citizen's initiative is not a new one. There is the well-known example of Henri Dunant whose actions, after he had seen 40,000 soldiers left dead or dying on the battle field at Solferino in 1859, led to the formation of the International Committee of the Red Cross [22]. Other international institutions that owe their existence to the persistent efforts of citizens include the International Criminal Court [23]. A parallel initiative today is the current project to create an International Court for the Environment [24].

This project will be part of a wider movement to develop ways for humanity to live in a sustainable relationship with the natural world of which we are part. Earth Jurisprudence, for example, is an emerging set of principles reflecting the change in our understanding of our relationship with nature, from an anthropocentric world view to one that sees humans as an integral and inseparable part of the earth system. The term 'wild law' is used to describe measures, such as constitutions and regulations, that give practical effect to principles of Earth Jurisprudence [25].

- New Zealand has passed laws that value the Earth for its intrinsic value.
- In 2007 the Ecuadorian government offered to forego drilling for oil in the Yasuni rainforest, one of the most biodiverse places on Earth, if the international community agreed to reimburse it for at least $3.6 billion over the next 13 years, or half of what it would earn from exploiting the oil. In August 2010 Ecuador and the United Nations Development Program signed an agreement to put in place a trust fund where countries can contribute to the initiative. Germany, Belgium, Italy, and Spain have already offered their financial support to Ecuador, and China, Korea and Japan have recently expressed interest in participating.
- Equador's new constitution has a chapter on Rights for Nature that creates a new regime of environmental protection.
- Bolivia proposes that the United Nations adopts a Universal Declaration of the Rights of Mother Earth, in terms similar to the Universal Declaration of Planetary Rights proposed by English lawyer Polly Higgins in her address to the United Nations in November 2008.

A first step: establishing a Global Climate Commons Trust

A first step towards establishing the new structure could be for citizens to establish a new independent organisation to implement and administer climate policies requiring some form of global administration. This would be designed to avoid the flaws in the UNFCCC system – for example it would be charged with acting on behalf of humanity as a whole, including future generations. Such an organisation would set specific global science-based targets, whether in terms of concentrations of particular gases in the atmosphere, or of reductions in world emissions into the atmosphere of particular gases, or of increases in draw-downs to take greenhouse gases out of the atmosphere into sinks and biomass. It would have the capacity to design and administer global policies that are both effective to meet the science based targets and also socially just. States would be invited to endorse and legitimise its operations on their own territories.

In discussion about this idea, the new organisation has been referred to as a Global Climate Commons Trust. The Trust would be established in a particular country whose laws include the centuries old concept of the trust, for example England or Ireland. Here we already have a basic framework of law within which trusts operate; plus the appropriate regulatory and court system. The law requires trustees to act with undivided loyalty to the purposes of the trust and they must act transparently. The Trust would be a recognised legal entity able to establish relationships with other entities including states; and obligations written into the constitution of the Trust to ensure transparency and accountability would be enforceable in courts of law.

The Trust could be given an initial remit to introduce and administer a global Cap and Share scheme, as described elsewhere in this volume. Its constitution could allow for other functions to be added later. Its first actions would be to set a global upstream cap on fossil fuels, issue the number of upstream permits limited by the cap, ensure that full market value is paid for these and arrange for the distribution of the proceeds of the permits – Richard Douthwaite"s chapter refers to five purposes for which these proceeds could be applied.

The Trust would need to comply with the principles of natural justice and the rule of law, for example, not purporting to set a cap retrospectively but giving reasonable notice of changes; giving stakeholders a chance to be heard before a decision is made, presumably, in view of the numbers involved, via some representative organisation or group; and not favouring any one supplier over others. A requirement to observe the rules of natural justice and to comply with the rule of law could be written into the constitution of the Trust; these duties could be enforced through the courts.

The legitimacy of the new structure and its relationship with the UNFCCC

Questions in readers' minds are likely to relate to the 'legitimacy' of the structure we are envisaging and its relationship with the UNFCCC and nation-state governments generally.

The 'legitimacy' of a Global Climate Commons Trust

Complying with the law, observing the rules of natural justice and the 'rule of law' do not of themselves confer 'legitimacy', an expression which has a wider and less well defined meaning. The practical test of legitimacy is general acceptance [26]. A Global Climate Commons Trust would be able to claim a tentative kind of legitimacy from the outset on the grounds that it constitutes a reasonable initiative to provide an effective way of addressing the concerns of the public about climate change and responding to the evidence of climate scientists, given the failure of current processes to address this grave danger effectively. Whether that claim comes to be accepted would be likely to depend on whether this institution succeeded in winning public support and the collaboration of nation state governments, bearing always in mind that noone will support, and no government will collaborate, unless they have chosen to do so. If and when the new arrangements had succeeded in winning public acceptance, they would then be 'legitimate'. That they may not be, indeed cannot be, so at the start is no ground for not starting!

Relationship with the UNFCCC

As to a Global Climate Trust's relationship with the UNFCCC, the Trust could be incorporated into the UNFCCC regime to take effect after 2012. The ultimate objective of the Convention, stated in Article 2 is "to achieve, in accordance with the relevant provisions of the Convention, stabilization of greenhouse gas concentrations in the atmosphere at a level that would prevent dangerous anthropogenic interference with the climate system. Such a level should be achieved within a time-frame sufficient to allow ecosystems to adapt naturally to climate change, to ensure that food production is not threatened and to enable economic development to proceed in a sustainable manner." The first principle of the Convention, stated in Article 3.1, is that the states signing the Convention "should protect the climate system for the benefit of present and future generations of humankind, on the basis of equity, and in accordance with their common but differentiated responsibilities". The new structure I am proposing, which a Global Commons Climate Trust would be part of, would provide a structure to enable states to help humanity implement this principle. The words of the Convention recognise that climate change is not primarily an issue between states: it is primarily an issue for the whole of humanity. But whilst incorporation into the UNFCCC regime to take effect after 2012 is a possible development, and could be very desirable, it is important to stress that the formation of a Global Climate Commons Trust is not necessarily dependent on adoption within the UNFCCC system.

Other likely developments

Establishing a Global Climate Commons Trust with an initial remit to introduce a global Cap and Share scheme would lead naturally to the development of other components of a new global climate governance system.

- I have referred above to the need to establish an independent worldwide climate science information service as part of the new structure. This would be needed to ensure that the decisions of the Trust are truly science-based.
- Every component of the new system would benefit greatly from a clear statement of its purpose [27]. In contrast to the current system's commitment to economic growth in order to maintain the debt-money system I have referred to the need for the new system to have the Planet "at the centre of rational loyalty for all human kind". The new system would need to have climate safety and social justice at its heart. It would benefit greatly from an express statement of the principles that would be applied at every level of the system, a Climate Commons Charter. A wonderful statement on which to base this already exists in the form of the Bolivian Universal Declaration of the Rights of Mother Earth [28].
- The Trust's arrangements for the distribution of the net proceeds of the sale of permits would need to be designed based on principles such as subsidiarity and the Viable Systems Model.

Misconceived? Or just a pipe-dream?

Doubts about this proposal, raised by people who are well aware of the climate crisis and the need to take action to reduce CO_2 emissions, are roughly of two kinds, or a mixture of the two. The first type of objection is that what is required is action by governments; the continuing rise in CO_2 emissions is driven by corporate imperatives and corporate powers that only governments can restrain. It is not something ordinary citizens should attempt to tackle themselves. The second point of view is that whilst this proposal is well-intentioned it is unrealistic to hope that it stands any chance of succeeding in the real world, so we would be wasting our time trying.

Objection (1): this task is for governments

In reply to the first type of objection, the key point is that climate change calls for international action. If the task of reducing total global emissions is left to the governments of nation states acting separately, and to the initiatives of citizens and cities, corporations and NGOs, there can be no certainty of achieving the necessary reductions anything like quickly enough. It is a task that requires urgent coordinated action by all nation state governments and other players. The current problem is that the process set up in 1992 to bring about the necessary coordinated action has failed. There is an empty seat behind the global government desk [29]. And the many initiatives outside the UNFCCC process are not coordinated. The question now for everyone is: how can the necessary coordination of nation state governments and other players be obtained? Our proposal offers an answer to that question: instead of relying on the necessary coordination to emerge from negotiations towards an agreement between nation-state governments, we propose a process whereby a judgement based on the latest climate science is arrived at by an independent body representing life on Earth and which also offers to administer worldwide schemes which will achieve the necessary targets if nation-state governments agree to give their officials the necessary instructions and/or corporations decide to cooperate voluntarily.

The proposal should thus be seen not simply as a new institution claiming to exercise powers properly the function of governments; but rather as a way of putting in place an appropriate and effective process to operate in lieu of the current failed process. The proposal does not challenge the sovereign powers of nation-states or the functions of governments in relation to their boundaries or economies. The legally binding nature of any scheme introduced by the new institutions will depend of the cooperation of nation state governments exercising control over their own territories. Nor is it being suggested that an institution should be created that will itself have powers of compulsion over oil and gas or coal-mining companies which chose not to cooperate.

Objection(2) this is just wishful thinking

The second type of doubt – that whilst this proposal is well-intentioned it is unrealistic to hope that it stands any chance of succeeding in the real world – is more a matter of attitude than of logic. In view of the information coming from top climate scientists and the record of governments in response, humanity undoubtedly faces a hugely serious crisis which governments are failing to meet; and noone has so far suggested any other way of making sure that greenhouse gas emissions and concentrations in the atmosphere are brought down fast enough. It is a situation that demands a new initiative. Whether this initiative succeeds will depend on whether it attracts enough support and participation, from distinguished leaders, from climate experts, from constitutional draftsmen, from publicists, above all from many millions people around the world. With the help of modern communications technology it is not unreasonable to believe that the necessary support and participation will be forthcoming and that people will be able to bring enough pressure to bear on their own government to persuade it to work with the new system.

True, most governments, and most government officials and politicians, can be expected to oppose the whole idea, since the justification for it is the failure of the existing inter-governmental system, and because it seeks to introduce a new extra-governmental actor into an area that has been assumed to be the sole responsibility governments. Some elements of the business world may be more sympathetic but a vast army of coal, oil and gas industry lobbyist are certain to support government opposition. The media cannot be relied on to be any more helpful.

Even if a Global Climate Commons Trust won widespread public support, it does not follow that governments would agree to cooperate. They, or the majority of them, will still be part of the Global Monetocracy with all the limits this implies on the possibilities open to them. Just as there was very widespread opposition at least in Europe to the invasion of Iraq in 2003, for example, this did not prevent the United States and Great Britain from going ahead with an illegal war. This is the kind of behaviour we can expect from powerful governments in the interests of multinational energy companies.

So how could the inevitable opposition be overcome?

It may partly depend on how the task of building a new system is begun. One possibility is that a Global Climate Commons Trust is set up by people of the highest standing, a group well-qualified to take on this responsibility. The same could apply to the proposed independent climate science information service. It would be possible to make sure that both are properly established so as to command respect.

A completely opposite approach would be to create an alliance between the people with the least standing in the world, indigenous peoples, around the rights of Mother Earth and the need for a global trust organisation to defend Mother Earth; and then, and only then, get "people of standing" to endorse the initiative of indigenous peoples.

One way to start would be to set about creating an "Upstream Fossil Fuel Data Base" to identify where coal, oil and gas is coming out of the ground and into the global economy – the locations and installations that will have to be rapidly closed down. Creating such a data base wikipedia style could be, indeed it would have to be, one of the first projects of an independent process. It would bring together people keen to support the idea of building a new system from the bottom up. It could be a project for people taking part in the Occupy movement.

Whichever way it achieves 'lift-off', to grow the initiative will have to attract widespread support from a wide variety of groups and interests. There must be a good chance of its doing so especially in the aftermath of Copenhagen, Cancun and Durban. The obvious incompetence of governments in managing the global financial system can only help.

The first and possibly the main tool at our disposal for overcoming the expected opposition will be direct campaigning. This will build on the massive campaigns involving millions of people conducted since the threat of climate change became a public issue in the 1980s. But it will transform the nature of these campaigns from that of trying to influence the way the current defective system works, essentially trying to persuade it to do things it is not designed to do, to that of inviting governments to collaborate with a new effective system, by performing the crucial, but not too difficult, job of policing the introduction of fossil fuels into their economies by banning the introduction of fuels not covered by a Trust permit. Moreover, the pressure on governments not to collaborate to be expected from the fossil fuels industry can be countered by pressure from the renewable energy industries which would benefit from the countries they operate in joining the scheme.

A possible positive knock-on effect may be to lift people out of their current condition of widespread denial and apathy, which although due mainly to ignorance and misinformation, is partly attributable to the lack of an alternative to the existing system. If truthful climate science information is accompanied by the practical possibility of taking part in building an alternative, people may be more ready to face up to the real danger we are in. Once the process of building the proposed new institutional arrangements has begun, this will provide everyone with something positive to do towards addressing the global peril, not just in their back yard and their local communities but at the global level. As well as being the best way of achieving the necessary reduction in fossil fuel emissions, by providing an alternative to the flawed international negotiation process, our proposal may indirectly help to unstick the general public denial – the general public's failure to recognise both the peril we are in and the incompetence of governments to deal with it. It attempts at any rate to bypass all these blockages.

Realising widespread support

The case I am making here is that the system the world needs in order to address the global problem of climate change must be global in scope, but not hierarchical (that is to say 'command and control') in character. The participants should be as diverse as possible. What will hold the system together is not top-down authority, or a compromise agreement thrashed out over years of negotiation between politicians and civil servants (as was the intention of the UNFCCC process), but a clear ethos shared throughout the system by everyone engaged in it. If we are going to establish a world-wide architecture to manage schemes like Cap and Share, in a highly complex, ever-changing, and, for many, dangerous context, it is essential that the participants can work together, trust each other, resolve differences, cooperate and coordinate and so on. The system we are establishing will have to be as diverse as the environment it is dealing with (Ashby's law), but if those engaged in it do not have shared values, shared 'heart', it is common sense to predict that it will fall apart – especially given the prospect we face of turbulent times ahead.

The proposition put forward here is that the world needs a system of managing human impact on the Earth's climate that does have widespread support and participation – unlike the present one. The project of creating such a system will stand or fall depending on whether it does attract widespread support. A chicken and egg situation.

The system must therefore be created and run by people who share the same values. Just how these are defined is not for consideration here. The wording can be agreed through a participative process and incorporated in a Climate Commons Charter; but the substance of the shared values cannot be negotiated. This initiative can only take off if there are people sharing the same values around the world to make it happen. There are grounds for confidence that there are. The participants in this task will be the millions of people around the world who understand the peril we are in and who put their responsibilities as world citizens ahead of other loyalties and interests. They are the worldwide community of ecologically engaged citizens. If we were not confident that this community already exists we could not even start to work on designing the new system.

As I have observed, in today's mainstream world the non-contestable objectives are those of economic growth and 'development'. But this initiative is not a product of the mainstream. It will be carried out by people who have given up any hope they ever had of responsible behaviour coming from the mainstream system. They may have reached that conclusion from personal experience, and from reading books by writers who have stood above the crowd, seen the current mainstream system for what it is and enlightened the rest of us. [30] These people have realised that the existing system identifies with the dominant elite, with Man as above nature entitled to exploit it as a resource without regard to natural limits, or to the importance of biodiversity and the health of ecosystems, or to the operation of the feedback loops that have kept the Gaian system as a whole balanced in a state favourable to human life for thousands of years. These people share a longing for the world to be ruled by a very different value system.

These values represent an understanding of life shared by millions and for thousands of years. They have even, over the last half century, been fighting their way to the surface of international relations, emerging in the form of UN declarations such as the Declaration of Human Rights, the Earth Charter, Biodiversity Convention and the Declaration of the Rights of Indigenous people. These charters and conventions, widely disregarded by governments when it suits them to do so, reflect values which are widely held throughout the world by individuals and non-governmental organisations, without whose persistent efforts many of these instruments would never have come into existence. They reflect the writings of thinkers, too numerous to mention, who have spoken out about our nature as social beings, and as part of, not separate from, the natural world.

These values are the values practiced in their lives by thousands of groups as identified, for example, by Paul Hawken's *The Blessed Unrest* [31]. They are shared by many now taking part in the Occupy movement.

An important consequence of the need for the new system to have a clear ethos and identity is that corporations constituted to put shareholder value above all other considerations cannot be part of it. They will continue of course with their business as usual; and they will no doubt continue to exercise influence over governments, but unless they change their dominant purpose to that of the public good, they cannot participate in the operation of the new system. By excluding corporate interests from participation in or influence over the new system, it will be free from one of the most disabling weaknesses of the current governmental system.

Taking part in a natural process

The task we are envisaging is massive in scale and difficulty. It may help to see that the process is a perfectly natural one. Here are four concepts about the way natural systems work that we can relate it to.

Growing up

Following birth and infancy, living systems pass through a competitive, self-interested, irresponsible, growth stage before this gives way to a mature cooperative stage of adulthood. During the growth phase of human civilisation, it was not surprising that a system of competitive markets developed, with the main players being corporations legally bound to promote their own interests above all other possible considerations and so contribute to growth, with a bank-created-debt money system that has to grow in order to avoid collapse, all legitimised by nation state governments largely unable to resist the pressure of corporate lobbying. The competitive adolescent growth phase cannot continue any longer in a finite world; and we have already come up against, and indeed passed, numerous limits. It is time humanity moved into its mature phase. Today we need a cooperative system of economics and governance designed to achieve a stable, steady state economy [32].

Adaptive cycles

Ecologists have suggested that all ecosystems pass through four phases – rapid growth, conservation, release, reorganisation; and then the same four phases again, rapid growth, conservation, release, reorganisation [33]. We can apply this model to the political process over the last few centuries. Where are we today in the cycle? Over the last century governments have got bigger and bigger. Societies today are ruled by countless pages of legislation and regulation administered by vast bureaucracies. Yet all this government has failed to bring us either peace, or social justice or sustainability. In particular it has spent the last twenty years failing to bring about a reduction in CO_2

emissions into the atmosphere, arguably the most important single thing governments should have done for the safety of people. Governments, and the growth economy that governments promote at all costs, have led the world into an extremely unstable state, in terms of both finance and climate. So we have obviously been enjoying, if that is the right word, the conservation phase. If analysts like David Korowicz and Stoneleigh [34] are right we must be getting very close to the release phase.

The significance of this is that this is the point when we can no longer take the governmental system as a given, as has been the case in most western nations through most of the 20th Century – the differences being fought out were generally about who should be in charge and what should their policies be, whether to have more or less state control and so forth, not as to the nature of government itself. Now, as we approach the reorganisation phase of the cycle, this is the time when we have to stand back and think hard about how we can reorganise. In the re-organisation phase many of the components of the existing system are put together in a new way, reconfigured to create the reorganised system. The discussion we are having in this chapter is timely in terms of the natural adaptive cycle of political institutions.

Creativity

The reorganisation phase of the adaptive cycle is where nature's capacity for creativity is realised. In *Creative Leaps Shape the World* William Graham Smith wrote "A very remarkable feature of the Universe is its capacity to create new kinds of things with new qualities, and hence to give rise to new overall situations, out of pre-existing ones.....There is now an urgent need for us to make a careful and sustained attempt to try to understand the nature of the problems that currently confront us and how we might use the creative potential existing on this planet, and not least within ourselves, to create a more satisfactory order of things here on the planet on which we live" [35].

Evolutionary leap

Complex adaptive systems, like the Earth, or governmental systems, rarely act instantly, they take some time to react. In systems-speak this is called system delay, inertia or lag. The Earth's climate has been slow to respond to the much higher than usual concentration of CO_2 in the atmosphere due to man-made emissions. Governments have been slow to respond to the looming climate change crisis. And we the public have been slow to respond to the failure of the current governmental system to cope with a wide range of problems, including climate change. Sooner or later, on all these fronts, change may start to happen very fast. Eckhart Tolle has written "When faced with a radical crisis, when the old way of being in the world, of interacting with each other and with the realm of nature doesn't work any more, when survival is threatened by seemingly insurmountable problems, an individual human – or a species – will either die or become extinct or rise above their limitations with an evolutionary leap. This is the state of humanity now, and this is its challenge" [36]. Will our species rise above the limitation of having inherited a useless governmental system? And will this happen before the Earth system slides into rapid and unstoppable change?

Those four concepts enable us to take on the task of building a new system for addressing climate change knowing that in doing so we are taking part in a natural process. These concepts are not mere metaphors. They help us to understand the nature of the natural world and how it adapts and evolves. Human social systems, human civilisation itself, are natural phenomena. They seem to share many of the characteristics scientists have observed in non-human living systems; and to pass through the same cycles.

Most major changes in human affairs, such as the Industrial Revolution, happen spontaneously. Today major changes of some sort are absolutely bound to result from oil peak and climate change during the next few decades. The only uncertainty is as to the nature and degree and timing of the changes. Under 'business as usual', a continuation of the growth economy, which is where current governmental systems are taking us, the future will be very ugly indeed. But we now know and understand enough to help bring about a conscious, deliberate revolution in the way we govern ourselves. We can start the process by constituting a Global Climate Commons Trust [37].

End Notes

1. James Hansen *Storms of my Grandchildren* Bloomsbury 2009; Clive Hamilton *Requiem for a Species* Earthscan 2010. Both authors are very worried not only about the science but also about the failure of politicians to listen to climate scientists. The chapter in which Hamilton describes the meeting of scientists he attended in Oxford ends with a reference to the international event in Copenhagen in December of the same year "Alas, three months later in the Danish capital those in command of the facts were drowned out by industry lobbyists and ignored by timorous politicians".
2. A classic framing of the problem is Andrew Hurrell's question: "Can a fragmented and often highly conflictual political system made up of over 170 sovereign states and numerous other actors achieve the high (and historically unprecedented) levels of cooperation and policy coordination needed to manage environmental problems on a global scale?" See Bulkeley and Newell *Governing Climate Change* Routledge 2010 page 5, quoting Hurrell *International Politics of the Environment* Clarendon Press 1992.
3. This was, for sample, the expressly stated objective of the UNFCCC's Executive Secretary Christiana Figueres, in her appeal to governments preparing for the Conference of the Parties held at Cancun at the end of 2010 http://unfccc.int/files/press/statements/application/pdf/pre_cop16_address_cf.pdf.
4. Bulkeley and Newell *Governing Climate Change* Routledge 2010 page 3.
5. "the IPPC's work has been heavily politicised from the very outset": Bulkeley and Newell *Governing Climate Change* Routledge 2010 at page 27. See also Fred Pearce, *The Climate Files: The battle for the truth about global warming*, 2010, Guardian Books, London; and Halina Ward *The Future of Democracy in the Face of Climate Change* Foundation for Democracy and Sustainable Development.

6. The incompetence of the nation state system has been described by both former politicians and academics: an example by a former British Foreign Secretary in a government led by Margaret Thatcher is Douglas Hurd's *The Search for Peace* Warner Books 1997. Philip Allott's *The Health of Nations, Society and Law beyond the State* Cambridge 2002 is the work of a Professor of International Public Law at the University of Cambridge who was formerly a Legal Counsellor in the British Foreign and Commonwealth Office.
7. In *Macrowikinomics,* Portfolio 2010 Don Tapscott and Anthony D. Williams describe how thousands of groups around the world are using the power of collaborative innovation and open systems to do something about carbon emissions. In *Climate Governance at the Crossroads,* Oxford 2010 Matthew J Hoffman identifies and analyses 58 'experiments' in climate governance, noting who initiated them and how they worked: market or regulatory, for mitigation or adaption, voluntary or mandatory; what activities were undertaken, for example: cataloging emissions, setting targets, education, monitoring, these being categorised as either planning, networking, direct action or oversight; the actors themselves being categorised as either networkers, infrastructure builders, voluntary actors or a mixture of all three. In a paper entitled *The Transnational Regime Complex for Climate Change* Kenneth W Abbott refers to the "proliferation of organisations, rules, implementation mechanisms, financial arrangements and operational activities" making up the regime complex for climate change governance, and maps these according to the identity of their constituent actors – from business firms to city governments to varied combinations of public and private stakeholders. see http://ssrn.com/abstract=1813198
8. For a recent description of the current system of creating money see article by Darius Guppy at - http://www.independent.co.uk/news/business/comment/darius-guppy-growth--it-aint-happening-2295967.html];
9. Madron and Jopling *Gaian Democracies redefining globalisation and people-power* Green Books 2003 pages 67-98.
10. Sian Sullivan *The environmentality of 'Earth Incorporated': on contemporary primitive accumulation and the financialisation of environmental conservation* Paper presented at An Environmental History of Neoliberalism, Lund University, 6-8 May 2010.
11. Most of the problems that governments have to deal with, including climate change, are what Prof Horst Rittel called 'wicked problems', meaning problems arising from non-linear complexities, such as 'the drug problem' or climate change, as opposed to 'tame' problems which arise from faults in linear systems, for example a machine. Command and control forms of governance are generally incapable of bringing about improvements in relation to this type of problem, as indeed the history of government in relation to climate change illustrates: such problems can only be addressed effectively through participative processes involving the 'stakeholders" in the problem. see Madron and Jopling *Gaian Democracies* Green Books 2003 page 41.
12. In *Storms of my Grandchildren* James Hansen expresses the frustration of top-scientists at not being listened to by governments dominated by corporate lobbyists. See also the *Democracy Now* interview with Hansen on 22 December 2009.
13. For example Michael Zurn, Director at the Science Centre Berlin, anticipates growth in the demand for new types of supranational and transnational institutions: see Held and Koenig-Archibugi Eds. Global Governance and Public Accountability Blackwell 2005 chapter 7 *Global Governance and Legitimacy Problems* at page 146.
14. Clive Archer, Research Professor at Manchester Metropolitan University sums up the emerging position: "A world in which mankind decides to confront universal problems by the use of effective international organisations will see a shift in the balance of political activity from the sovereign state to a number of strengthened globally functional (but also highly political) institutions". *International Organisations* Routledge 2001. See also Phillippe Sands *Principles of International Law,* Cambridge 2003 2nd Ed pages 11-18; and Joseph A. Camilleri and Jim Falk *Worlds in Transition Evolving Governance Across a Stressed Planet* Edward Elgar 2009 pages 551-563. In *The Health of Nations – Society and Law beyond the Nation State*

Cambridge 2002 Philip Allott describes the "revolution in our minds" needed to "leave us free to make and remake a human society which does not abolish our national societies but embraces and completes them".

15. EF Schumacher *Small is Beautiful* p 228.
16. Stafford Beer *Designing Freedom* Wiley 1974, *Diagnosing the System* Malik 2008 and *The Heart of Enterprise* Malik 2008; for Jon Walker's VSM Guide see http://www.esrad.org.uk/resources/vsmg_3/screen.php?page=preface
17. Stafford Beer, *Designing Freedom* Wiley 1974
18. In *Global Governance and Legitimacy Problems* (chapter 7 of *Global Governance and Public Accountability* Held and Koenig-Archibugi Eds.) page 162, Michael Zurn points out at that denationalised governance structures are both good for democracy and for giving back to national policy makers the capacity to deal effectively with de-nationalised economic structures.
19. Joel Bakan *Corporation* Constable 2004; Marjorie Kelly *The Divine Right of Capital.* Berrett Koehler 2001
20. Jeremy Leggett *The Carbon War.* Routledge 2001
21. Steve Waddell, for example, has developed the concept of GANS, or Global Action Networks, illustrated by organisations like Transparency International and the Forest Stewardship Council. Waddell has identified five elements common to the strategies used by these networks. http://www.capacity.org/capacity/opencms/en/topics/multi-actor-engagement/global-action-networks
22. http://en.wikipedia.org/wiki/International_Committee_of_the_Red_Cross
23. http://www.icc-cpi.int.
24. www.environmentcourt.com.
25. See www.ukela.org and www.yesmagazine.org/planet/advice-for-water-warriors.
26. Compare the observations of the late Lord Bingham in *The Rule of Law* Allen Lane 2010 on the significance of general acceptance as the basis of international law (at p113) and of long acceptance as the basis of the principle of the sovereignty of Parliament in the United Kingdom (at p167).
27. See the statement by Dee Hock at http://en.wikipedia.org/wiki/Dee_Hock#Quotes; and the role of System 5 in Stafford Beer's VSM for which see [16] above
28. http://pwccc.wordpress.com/2010/02/07/draft-universal-declaration-of-the-rights-of-mother-earth-2/
29. In his Democracy Now interview in December 2009 James Hansen remarked "I'm actually quite pleased with what happened at Copenhagen, because now we have basically a blank slate".
30. Writers who have opened the eyes of this writer include Susan George, David Korten, John Pilger, Noam Chomsky, Melanie Klein, Michael Rowbotham, John McMurtry, Vandana Shiva, Mark Curtis and Eduardo Galeano.
31. http://www.youtube.com/watch?v=N1fiubmOqH4]
32. Elisabet Sahtouris video After Darwin Pt 2 http://www.youtube.com/watch?v=j_immL2m1tg&feature=related
33. Gunderson and Holling Eds *Panarchy* Island Press 2002
34. See www.feasta.org and http://theautomaticearth.blogspot.com/
35. William Graham Smith *Creative Leaps Shape the World* International Books 1997
36. www.eckharttollecontactsuk.org/Biography.html
37. For suggested first steps, and also a snapshot of what the world might look like in 2015 if a Global Climate Commons Trust had been established, see www.capandshare.org/climatecommonstrust

Chapter 6

Cap and Share in India

James Bruges

Should the funds from Cap and Share be distributed equally to individuals or are there better ways of using them? Indians are even more in need of financial help than those suffering from austerity programmes in the west, but I give reasons why it would be better to distribute to community organisations in India. I will start with some comments on climate and also cover related economic issues.

A changing climate could lead to suffering and death on an unimaginable scale, so let's think of our use of fossil fuels as a disease, a cancerous tumour that must be removed from society if it is to survive.

The rising cost of these fuels is one of the reasons the economy has been destabilised, but which is more important, a stable economy or survival? Perhaps we can have both if we rely only on renewable energy and change the monetary system. For any policy-maker with an ounce of intelligence this should have been obvious for a hundred years, as is suggested by this statement: "The unexpected legacy of fossil fuels leads us to lose sight of the principle of a durable economy, which needs to be based exclusively on the regular influx of energy from the sun's radiation." It was said by Wilhelm Ostwalt who received the Nobel prize in Chemistry in 1909. His statement is a stark example of *homo sapiens'* ability to ignore inconvenient truths.

Cap and Share was developed by Feasta (Foundation for the Economics of Sustainability), and is a radical and realistic policy for weaning us off fossil fuels within a defined period.

If applied globally, Cap and Share would require a United Nations agency to put a ceiling on the amount of coal, gas and oil that can be extracted from the ground each year, gradually reducing to zero. All companies that mine these fuels would need to buy emission allocations from the agency before selling their fuels. The agency would distribute the funds raised to a Trust in each country in proportion to its adult population. These national Trusts would only be recognised by the UN if their policies relate directly to issues of climate and poverty. A quarter of the funds generated would be retained to finance measures that benefit those that are not yet adult.

Income from the sale of allocations is likely to increase as demand for increasingly scare fuels rises, though the income will not rise steadily until the fuels run out. When petrol, for example, is no longer the most common fuel for vehicles it will become increasingly uneconomical to maintain a network of petrol pumps, and increasingly difficult to sell cars that are made to run on petrol. So the income from petrol will rise until it reaches a plateau where people can't afford it and alternative fuels take over, after which there will be a sudden collapse of income. The same pattern will develop for the heating of buildings using gas-fired boilers and for the use of energy-intensive fertiliser for agriculture.

So there will be a tipping point with a wholesale shift from fossil sources to renewable sources for energy. There are wildly different guesses for when the tipping point will be reached but it could be in 15 to 20 years. The funds raised by Cap and Share, therefore, would not be a steadily increasing stream, but would rise and then suddenly collapse.

Such a global agreement seems wishful thinking in the present state of global politics. I'm also becoming sceptical about the possibility of overarching policies to solve the world's problems. The only time heads of state agreed on climate issues was at the Rio Earth Summit in 1992. In the two decades since then lobby groups have managed to prevent their agenda being implemented. It seems that failure is due to 'bigness', in parallel with failure of the various 'isms' – communism and capitalism – that advocates have tried to impose at a global scale.

Cap and Share can be applied within a country unilaterally in the absence of a global agreement. At present two regional versions have been devised for developed countries. As formulated in Ireland, all adults would be given an equal number of emission coupons, reducing each year, for sale to companies that mine or import fossil fuels. Cap and Dividend is a similar proposal for the USA where emission allowances would be auctioned to fuel companies and the proceeds would be distributed to all citizens equally (a similar policy already exists in Alaska with the distribution of income from oil extraction to all citizens). Both these approaches are designed to achieve the surgery – removing the fossil-fuel tumour – that is Cap and Share's primary objective.

India could also act unilaterally. It could set a cap on the use of fossil fuels and establish a Trust that would raise funds by auctioning allocations to producers and importers of these fuels within its borders: so far the same as for developed countries. It is the share that should be handled differently. Anandi Sharan is an economist who carried out a study of Cap and Share for India in 2008 to assess the level of funding that might be raised. Since then the concept has had exposure in the country. For the share, she considers that India already has an appropriate mechanism for helping the poor majority.

The Indian elite is euphoric about its economic growth, though spiralling inequality is likely to land them with catastrophic problems, as is happening with us. The country has a long alternative history of seeking a sustainable future for the majority of its population through *panchayat raj*, a policy that devolves decision-making to the lowest level of community organisation. These are two opposing forces in Indian politics. The first – 'shining India' – aims to create wealth for the nation through the western economic model and is, nationally, the dominant political force. The other, based on Gandhi's ideas for local self-government, has a huge backing of activists, and is the dominant policy in some states. So should we be aiming at a 'monetary' Cap and Share as in developed countries, or should we work through policies that are already in use in India to help the poor? Sharan recommends the latter and I cannot help but agree. The Irish and US approaches are based on money and the individual. The *panchayat raj* approach is based on land and the community. *Panchayat raj* is particularly appropriate for rural areas that contain the majority of the country's population, mostly getting a living from the land, and mostly poor.

Under *panchayat raj*, the funds from Cap and Share would go to village and city groups (*gram sabhas* and *wards*) based on their population numbers. These groups, themselves, would decide how to manage the funds. It is an embodiment of the statement made by Gandhi in 1925, "Freedom is to be attained by educating the people to a sense of their capacity to regulate and control authority."

At this stage I had better mention my limited credentials for commenting. My childhood, up to the age of twelve, was spent in India. I have visited the subcontinent for the last fifteen years. My friends Stan and Mari Thekaekara have worked with *adivasis* (tribal people) for the whole of their adult lives. Mari also works with, and campaigns on behalf of, *dalit* latrine workers. Their AMS cooperative (also known as Just Change) is run by groups that produce essential commodities and enables its members to buy these commodities at the cost of production. This is just one example of the benefits of control by communities.

In 2002 I met Dr Cletus Babu, the founder of SCAD (*Social Change and Development*), a large NGO in Tamil Nadu that works with villages and small-scale farmers. In 2008 we discussed biochar: its role in tackling global warming and the benefit it might bring to the farmers with whom SCAD is in touch. He immediately saw its importance and asked if we could collaborate. With David Friese-Greene I am now involved with the Soil Fertility Project at SCAD, which uses biochar fertiliser to replace synthetic NPK, and increase the yield of crops on a permanent basis. Removal of carbon dioxide from the atmosphere is not the motivating force for the farmers. I mention this project because it demonstrates how lasting benefit can be achieved by small-scale farmers in ways other than receiving a handout of cash.

Incidentally, I asked Cletus whether SCAD could handle the distribution of Cap and Share funds to individuals, and he rejected the idea. It would distort the communal self-help philosophy he had followed over twenty-five years and SCAD would have to set up and manage an unwieldy bureaucracy in which the dangers of corruption and favouritism would be a constant source of anxiety. Due to his reaction I would question the possibility of finding independent and reliable organisations that could handle the distribution of funds.

I have also taken part in climate seminars in Kerala. It was at one of these that I explained Cap and Share and was severely shot down on the method we propose for the 'share'. They, unanimously, considered that the equal distribution of funds to individuals, over the 15-year duration of the policy, could be socially disastrous and provide little long-term benefit.

Reflecting on these experiences I assembled some thoughts on the subject, though these are little more than comments that any casual visitor to the subcontinent might make. I am particularly grateful to Anandi Sharan for discussing, correcting and providing further information.

Global climate negotiations are ongoing, but it seems unlikely that a binding UN protocol to limit emissions will be achieved anytime soon. However, a global Climate Commons Trust, set up by civil society, could promote the concept and seek support from countries that wish to take part (as mentioned earlier in this book). These countries could adopt separate initiatives – like the Irish approach for Europe, and Cap and Dividend in the US – and take measures to protect themselves from unfair competition.

Each country may have its own reasons for restricting its use of fossil fuels: it may have a principled awareness of the need for all large nations to play their part in tackling climate change, it may wish to wean itself off dependence on increasingly expensive imported fuels, it may wish to move its industries from technologies of the past to those of the future, it may be aware of the need to protect itself from tariffs and sanctions, or it may respond to a combination of several incentives.

Countries will be reluctant to reduce their own use of fossil fuels if this means their products suffer a comparative disadvantage. They should therefore be allowed to levy an import tax on carbon-intensive products in order to bring prices of imported goods into line with domestic ones. This tax would only apply to products from countries that do not restrict their own use of fossil fuels. This, of course, requires changes to relevant regulations imposed by the World Trade Organisation. The attempt by the US to prevent China subsidising its renewable energy technology indicates how important it is for climate policies to take precedence over international free-trade agreements.

In both the EU and the US versions of Cap and Share the 'share' aims to make the 'cap' acceptable to the public within the country through the distribution of money. But, as I found in the seminars mentioned above, there is a feeling that this would be wholly inappropriate in India. About half of all Indians are dependent on the land and on the public distribution system for subsidised food grains at Rs 1-3 (1-3 pence) per kilogram. For these people to benefit in the longer term, the money should be better used to improve the yield of their crops and achieve sources of energy and fertiliser that will survive the demise of fossil fuels. This would not be achieved if the funds were distributed direct to individuals.

A quarter of the country's population, 300 million, are *adivasis* (indigenous people) and forest dwellers. Many *adivasis* look to the mutual support of their tribe rather than to individual accumulation of wealth. I have met some of these tribes through the Thekaekaras and it is clear that the money-based economy of the modern state is disrupting their system of mutual support, and causing serious social problems.

The relatively recent imposition of western economic practice has resulted in India becoming one of the most unequal nations in the world. Graph after graph in *The Spirit Level*, a book by Richard Wilkinson and Kate Pickett, show how social problems in developed countries are directly related to the inequality within those countries. India has taken it one step further: the government is waging war on its own people in its pursuit of economic growth. I refer to the many forest people who are struggling to assert their rights and gain control of India's forests under the Forest Rights Act. The government is using the Forest Department to make the land available for mining and the steel industry and treating the forest people and their sympathisers as 'insurgents', thus forcing them into the hands of Maoists activists who give assistance that the government fails to provide. Having defined them in this way the government is now brutalising its own people in Chhatisgarh, Orissa and West Bengal. Binayak Sen, who had worked as a doctor extending health care to poor people in Chhatisgarh, advocating non-violent political engagement, was jailed for life for merely talking to 'insurgents'. This caused furious protests across the country and, following a ruling by the Supreme Court, he was released on bail.

Though the elite in India is steeped in commercialism and western economic values there remains a strong political movement for raising the quality of life of the poor majority.

Classical definitions refer to economy as "land, labour and capital". Taken in that order, the meaning comes close to traditional Indian values. 'Land', standing for the environment (earth, water, sky and living things), is the source of all wealth. This is particularly relevant to India where so much of the population is directly dependent on the land. 'Labour' refers to the wellbeing of society, and is the focus of socialist thinking that has had a big influence on Indian politics since independence. 'Capital' refers to the role of money – a social construct and a unit of account – which should take its place as an enabler, not as the source or the definition of wealth.

In Indian mythology, land is mother earth that has sacred trees and groves to which access is restricted. Mountains are the abode of gods, not challenges to be conquered. Linked to land's physical properties is the Indian attitude to animals. One finds the elephant-headed Ganesha – the charmed son of Shiva and Parvati – sitting on the dashboards of taxis giving luck (they need it!). In my photos of rural families a cow is almost always included in the group. An animal may have been a person in a previous life, and you may be re-born as an animal.

So 'land' – rather than money (capital) – holds the meaning of wealth in traditional Indian thinking. This may be why the seminars objected so strongly to our approach to the use of funds. A 15-year windfall income should be devoted to providing those things that the poor need from the land on a sustainable basis. True development must protect and enhance the properties of the earth, water, sky and biodiversity. People, particularly those directly dependent on the land, will derive long-term benefit particularly if the soil's fertility can be enhanced on a permanent basis, hence my interest in biochar.

Western attitudes to economics are so entrenched that I would like to comment on the reasons why they are causing harm. It is often mentioned, correctly, that both debt and growth are the foundations of our economy: 'this is how economies work, how they have given us prosperity. The rules cannot be changed.' They say! They admit, however, that it is a man-made construct dependent on confidence. By definition, therefore, the rules can be changed. And recent events indicate that they must be fundamentally changed. It is no longer possible to ignore evidence that the rules are unsustainable for economic activity, for society and for the environment.

To understand the centrality of debt it is necessary to understand how money is brought into circulation. A full description of this process is described in The New Economics Foundation's guide, *How Money is Created* (October 2011), and summarised on the Positive Money website. But I will attempt a layman's description. Banks do not just manage the money we use. They create it. Whenever they make a loan they bring money into existence. It is money that did not previously exist. If all businesses and people paid back the money they had borrowed from banks there would be no money in circulation and the economy would collapse. This is why, in a recession, we are not encouraged to save but to borrow and buy in order to get the economy back on its feet.

Then consider the profit the banks make from being allowed to create money. You take out a mortgage for £200,000 at 7% interest. At the end of its 10 years period you will have paid back the loan, but on top of that you will have paid an equal amount in interest. The only cost to the bank has been typing figures into a computer at, say, £5,000. So with £5000 outlay the bank will have made a profit of £195,000. No other organisation can match that as a business model! All the money that banks create attracts interest, which is why they are described as the lords of the universe living on a financial planet all of their own.

This gives a hint at the faults of the western monetarist model. One might just add a couple of comments by the governor of the Bank of England, Mervyn King: "It is hard to see why institutes whose failure cannot be contemplated should be in the private sector." And, "Of all the many ways of organising banking, the worst is the one we have today."

So much for debt. It is also called a growth economy because the economy has to grow in order to add interest to the money lent. This means that an increasing amount of resources have to be used. Metals have to be mined and more food has to be produced. It was the fashion among economists to question this during the 1995-2000 'dot-com' bubble that saw huge economic activity requiring little resource input. But after the bubble burst economists became too embarrassed to pursue this hypothesis. Most of the growth required by the system is for luxuries not essentials. We are urged to buy things we don't need with money we do not have. If our financial regulation requires growth then the finite resources of the world are threatened. Kenneth Boulding, economist, educator and Quaker, summed up the absurdity of a system dependent on growth: "Anyone who believes in indefinite growth in anything physical, on a physically finite planet, is either mad or an economist."

I find it interesting that passages in the first chapters of the Bible set out ideas for an economy. The two critical rules are an embargo on charging interest-on-money, and that debt should be cancelled every 50 years. This approach protects the debtor as well as the creditor. Ours only protects the creditor. It suggests to me that it could be morally and legally right to default on certain debts both by nations and by individuals.

The rural poor are not benefiting from 'shining India'. Debt is causing a epidemic of suicides, their land has been degraded by synthetic fertilisers, and the emphasis on industrial development is reducing land and making the country dependent on imported food, which the poor can't afford. We should not discourage India from the more holistic approach of the *panchayat raj*.

Labour and land in our system are treated as little more than financial assets to be bought and sold at the dictates of the market. Land, in particular. 70 percent of our land is owned by 0.3 percent of the population. The Duke of Buccleach owns 110,000 hectares, a situation incomprehensible in India where land distribution and a ceiling on land ownership are political benchmarks. Mari Thekaekara was staggered when she heard that one man could own so much land. She commented, "What does he do with it all? Does he farm it or even visit it? Does he know the people who live there?" Our legal system protects this large chunk of the nation's land as a source of income for an individual. John Stuart Mill made a relevant comment in 1848, "Landlords grow richer in their sleep without working, risking or economising."

The increase in the value of land, arising as it does from the efforts of an entire community, should belong to the community and not to the individual who might hold title. Bankers love our system, of course. They need to invest their millions somewhere, so it is in the interest of people with money that prime land in cities, for example, should constantly increase in value. In a recession the productive sector – like industry – is risky and receives minimal investment. When the banks were bailed out the majority of this money went into assets, like land, not into small businesses that would increase employment. The banks love secrecy so that money from Russian oligarchs and African kleptocrats via tax havens pushes up the value of assets ever further. Even agricultural land is subject to the same casino. Half the funding of the Conservative party comes from City sources, and three-quarters of the coalition cabinet are millionaires (Cameron £4m, Osborne £4.6m, Clegg £1.9m). Would it be impolite to suggest that this influences the government to act on advice from the very people that created the financial chaos from which we are all suffering? Why else does it allow the value of assets to spiral upwards? Why else does the government not channel funds direct to the productive sector rather than bailing out the banks? Why does it oppose the Tobin tax? The resultant dramatic increase in inequality is leading to widespread unrest though, hopefully, not up to the scale of the French Revolution. One might add that three-quarters of Indian ministers are multimillionaires.

We should learn from Indian philosophy that land is the source of all real wealth. Its rise in value is almost always created by decisions of the community – through, for example, planning laws, transport provision, public open space – but this rise in value goes to the individuals that exploit it. The obvious answer is a Land Value Tax, which could significantly reduce taxes on production (income tax and VAT) and bring us more in line with classical economics and Indian philosophy. For obvious reasons the banks and the cabinet oppose it.

Who should decide on the action to be taken in the use of funds generated by Cap and Share?

India has a highly developed concept of what we term subsidiarity. At one extreme is Britain's centralised government where little freedom is allowed even for local authorities. At the other extreme was Mahatma Gandhi's abiding concern with local self-government (*panchayat raj*) with decisions taken at village level (*gram sabha*) by all inhabitants, who would only refer to higher authorities matters that could not be decided at this level. Nehru thought *panchayat raj* was inimical to economic development and he was probably right, though his concept of development can be questioned. Gandhi's approach indicates a totally different concept of human development and this might have taken us down a route that would not have led to the present global crises. The *panchayat raj* policy persisted among some states and was revived by Rajiv Gandhi, finding its way into the constitution in 1993 after his death.

Panchayat raj in the constitution devolves significant powers to local institutions and allows the government to deal directly with them, bypassing State administrations if necessary. However the policy is managed pragmatically. States have a veto in the process, and can decide which decisions are taken at the local level, though the *panchayats* have responsibility over development work: irrigation, education, health, road-building etc. States, through the State Finance Commission, have the final say in how much money is devolved to this third tier of government. Kerala devolves over a third of plan-funds for programmes designed and executed by local institutions, while officials and technical experts help villagers to set their own priorities. This has resulted, in Kerala, in the careful management of natural resources such as soil, water and forests. Other states, such as neighbouring Karnataka, devolve practically no funds.

The Mahatma Gandhi National Rural Employment Guarantee Act, MGNREG, currently delivers around Rs40,000 (€660) a year to a family with four wage earners, and it distributes up to Rs150,000 (€2,500) per acre for land-based soil improvements, sapling and so on. It also guarantees 100 days of public works at around Rs80 (€1.30) per day (though this is below the minimum wage and should therefore be illegal). India is currently debating how to implement a welfare state as well as the development model for the next five-year plan. Anandi Sharan and her colleagues have also argued that India needs a Universal National Guanranteed Minimum Wage to ensure the transition to sustainable development happens without the minimal lifeline protection from the MGNREG Act being withdrawn.

Panchayat raj, therefore, provides a pragmatic approach allowing negotiation between the parties involved. The first port of call is the village council or town ward, where the law states that decisions should be by consensus. Then the *gram panchayat* of elected members covering several villages. Then *block and district panchayats*. Town municipalities are the next higher level for urban localities. It will take a revolution in India to implement the principle of subsidiarity since elites have captured the agencies of the state at state and Union level and will not let go. Though, on a positive note, the revolution has already started in Bihar, where an immensely popular *janata dal* (United) led by Nitish Kumar recently won a second term on a mandate of transparent clean development based on agriculture and land reform. What happens in Bihar could spread to other states.

Anandi Sharan's 2008 study found that, under a money-based Cap and Share, the annual earnings of a family in the lowest tenth of the population would rise from about Rs5,000 (€85) to Rs100,000 (€1,660) if the price of CO_2 reaches €100 per tonne. More if the price of CO_2 rises further. A domestic Cap and Share would obviously generate less income for the family. In recent discussions she rejects this individualistic approach and makes the important observation that the process of democracy would suffer through individualism but is enhanced by participation in the *panchayat* system.

Both the *panchayat raj* and the MGNREG indicate that India has an elected local government structure that is ready and available for the handling of funds generated by Cap and Share, whether from a global scheme or from the auction of emission allowances in India. This process determines which bodies should receive funds and for what purpose they should be used. Importantly, it is an existing structure that would be strengthened by Cap and Share and would remain after Cap and Share ceases to provide funds.

Most of the Indian population is dependent on the land. Until recently India has been self-sufficient in food but this is no longer the case. The government is making matters worse by giving priority to industry and the extraction of resources. Davinder Sharma reports that in one agricultural state, Uttar Pradesh, the government has designated over a quarter of its rich farming land for industrial use, and has also terminated the ceiling on land ownership. A majority of the farmers do not want to dispense with their ancestral land. Displaced farmers get a bonanza when their land is taken but how long will these funds last? Perhaps the farmers may get jobs in cities, perhaps not. The government argues that only industry can increase the nation's wealth and this will enable food to be bought from abroad. A recent study, however, found that the marginal cost of importing one tonne of corn costs $150, but only $30 if sourced domestically using fertilisers. And if world prices rise as forecast... Without land the displaced farmers will have to buy all the food they need rather than surviving largely on their own produce. Gone are the days when Jawaharlal Nehru, India's first Prime Minister, said: "It is very humiliating for any country to import food. So everything else can wait, but not agriculture."

The UN Food and Agriculture Organisation (FAO) is now warning of a food crisis that will lead to global scarcity and higher prices for imported food. This will be on top of food price inflation in India, which has been over 12% for two years. Rising fuel and fertiliser costs will continue to put pressure on food-production, and thus on consumer prices.

These issues of domestic policies and global shortages point to the urgent need to direct funds towards quasi-landless agricultural labourers, with a need for land reform and community managed sustainable agriculture. These needs are clearly understood by the *gram sabhas* that would determine policy under *panchayat raj*.

Another remarkable aspect of India is its acceptance of pluralism. For example, it allows all regional languages to be used, seventeen of which are printed on banknotes. This contrasts with neighbouring countries that tried to impose a single national language. The civil war in Sri Lanka was largely caused by the insistence on only Sinhala to be used while Tamil was outlawed. In Bangladesh Jinnah's insistence that Urdu should be the only language led to secession. Again this suggests a pragmatic, rather than dogmatic, approach where the use of funds should be targeted on policies that would benefit the majority in the long term with local needs in mind. It could be that *gram panchayats* and city wards would be responsible for administering part of the funds, and higher levels to administer climate issues that could not be managed at the local levels.

Cap and Share is largely concerned with reducing emissions. But there is also a need to reduce existing concentrations of carbon in the atmosphere. Biochar is the only 'natural' way to do this if we wish to avoid the unpredictability of geoengineering. I will finish by mentioning an initiative in India.

SCAD, *Social Change and Development*, is a non-governmental organisation in Tamil Nadu introducing the use of biochar to the thousands of farmers with whom it deals. The primary objective is to increase the yield of crops but it would also permanently lock carbon into the soil.

Biochar is charcoal, largely carbon, of a type that can improve the fertility of soil. All soils contain some charcoal left from forest fires over the centuries, and the use of charcoal in agriculture has a long history in India though it fell out of use with the introduction of synthetic fertilisers that provide plants with the main nutrients they need. Land is degraded when the interaction of plants and soil has been broken by synthetic fertilisers, and much of the land in SCAD's area is now described as semi-desert. These fertilisers are becoming increasingly expensive and government subsidy for them is reducing, so farmers are looking for other ways to maintain the yield of their crops.

SCAD became enthusiastic about charcoal when a banana grower in the area, Pattu Murugeshan, said that he had been using charcoal and ash from a rice mill for four years. His use of water had halved, his need for labour and fertilisers had reduced and the yield of bananas had increased by a third. His neighbours had been adopting the same practice.

SCAD's Soil Fertility Project is using biochar made from an invasive scrubland bush, prosopis, and from agricultural waste. Traditional charcoal making is avoided because of health hazards and its need to use wood. Instead, the pilot scheme has a robust charcoal pyrolyser that is being developed by BiG (Black is Green Pty) of Australia and now also being made under licence in India. The other main equipment is an anaerobic digester designed and made by Biotech in Kerala, the neighbouring state, which is run on green cooking waste from the central kitchen of a 5,000 student college, plus green waste from town markets.

Slurry from the digester is combined with biochar to make biochar fertiliser. While ordinary compost is of value to the soil, its benefits are temporary. With biochar fertiliser soil fertility increases on a permanent basis since the carbon structure of biochar provides refuge in cavities for moisture as well as for microbes, fungi and mycorrhizae that migrate from the digester slurry. An added benefit is that gas from the digester provides fuel for cooking and for generating electricity. The only input is waste plant material and labour. The pilot project includes rigorous scientific tests as well as field trials and some distribution to farmers and women's groups.

If the demonstrations are successful it is anticipated that demand will increase, villages will produce their own slurry and a mobile pyrolysis unit will process dry feedstock that they gather. The use of biochar fertiliser will at least partly replace synthetic fertilisers and, hopefully, the practice will spread among farmers.

This is, of course, a simplified description but it points to the possibility of soil enhancement and increased yield within a self-feeding cyclical process. Above all, it is a community project, which achieves benefits and strengthens the SCAD community.

Cap and Share would help in providing the necessary funds for research, for the initial introduction of appropriate equipment, for access to markets, for farmers' secure use of land (long-term improvements will not happen without this) and for encouraging the practice generally. The SCAD project is funded by a private charity and is carrying out its research with the help of CSIRO in Australia and the Universities of Limerick and Edinburgh.

These farmers are driven by the need to increase yield, not by a desire to extract carbon dioxide from the atmosphere. However, all plants capture CO_2 through photosynthesis, and when charred and buried the carbon remains in the soil. Small-scale farmers manage much of the world's productive land and the universal use of biochar by them could be a – possibly *the* – major contributor to reducing carbon dioxide in the atmosphere.

Glossary

Panchayat raj: traditional local assembly to settle disputes between individuals and villages. Recent governments have decentralised several administrative functions to the village level empowering elected *gram panchayats*.

Gram sabha: The quarterly meeting of the village *panchayat*, open to men and women over 18 years old, where all decisions of community development should take place.

Ward: administrative unit of a city.

Anandi Sharan, founder of Women for Sustainable Development, prepared 'Potential Impacts of a Global Cap and Share Scheme on India' in October 2010. She has submitted the following text concerning its application.

The Green Party of India is working for an all-party coalition government for nature and climate reconciliation. We were thus gratified when government politicians from the United Progressive Alliance actively addressed the management of the commons at a recent meeting in Delhi addressed by Elinor Ostrom, Nobel Memorial Prize in Economic Sciences (2009). We believe that the next five years will be absolutely critical, and the developments in Delhi give us some hope that we have the opportunity of working towards the formation of a government of emergency preparedness.

In the coalition government, the concept of the commons will replace the concept of economic growth. Practical day-to-day human political ecology – (can we define this as the day-to-day management of human affairs based on maintaining nature's bounty and being prepared for nature's wrath?) – becomes the guiding paradigm for the state.

The role of commercial companies is then to provide for basic needs: good water, good food, good clothes, good homes, affordable and reliable health care, and a good price for our labour so that all families can stay out of debt: this is what we will have at the forefront of our common endeavour. It will be facilitated by a frugal attitude, the correct price for energy based on internalising the cost of greenhouse gas permits to reduce competition with labour, and a scientific attitude to society. This will create the framework for climate and nature reconciliation. Cap and Share of greenhouse gases is part of this.

Cap and Share is a simple ecological tax reform to phase out fossil fuels and thus make possible the transition to adaptation to global warming. Though a global climate regime is unlikely, choosing a humane and spiritual national pathway of man-nature-climate-reconciliation will mean introducing Cap and Share unilaterally. Not only will this reduce global greenhouse gases, it will also support our human political ecology as we urgently strengthen the social systems to adapt to runaway climate change.

Managing the atmospheric commons through Cap and Share (the commons at *gram sabhas* level have similar management requirements which we have outlined in our manifesto) at a national level means to

- decide that doing it alone is in the interest of the country,
- create 1 billion permits (each permit is 1t CO_2e) in 2012, and auction them to the highest bidders. Only primary producers of oil, coal and gas can bid, and they are not allowed to dig out resources from the ground without such a permit and then only to the extent to which they have permits.
- From 2020 onwards monitor every year the sequestration of greenhouse gases in the country's forests and issue permits only to the extent of that sequestration capacity,
- have a national dialogue on priorities for spending the funds raised from the auctions: the prices of the permits are likely to be very high, but we only have 10 years of revenues. So the money must be used to create infrastructure for the third tier of government by means of untied grants to *gram sabhas* and town *wards* to allow decentralised planning for climate and nature resilience,
- create committees of political and social forces to usher in an all-party government of climate and nature reconciliation and manage and monitor funds: discuss the use of these untied funds for a national universal guaranteed minimum wage.

This basic agenda for an all-party national coalition for nature and climate reconciliation will thus accept the country as a natural system – a commons with boundaries – and will reassert the will to control prices and wages.

Note: Prices and wages are determined within a certain physical system. If as a government you open your system to the world, you lose control of prices and of money supply and of investments. The reports that government has no control over the present food price rise should not have come as a surprise to us. What we should not accept is that it is inevitably so, nor need we accept that wages inevitably cannot keep up. The all-party national coalition for climate and nature reconciliation will provide a real alternative to the global economic growth paradigm that leaves prices and wages to the market. If we bring in the universal guaranteed minimum wage financed from the revenues from Cap and Share and keep wages in line with food prices, we will be on a feasible and realistic and equitable survival pathway. We will manage the commons through Cap and Share of greenhouse gas pollution – the critical common resource that determines prices of energy and thus the level of wages.

This ecological tax reform will channel funds to the three-tier government structure which will allow local enterprises and agriculture to operate in the relatively unregulated manner we have become accustomed to – and as is necessary for climate and nature adaptation resilience. But because government is regulating the energy price, prices and wages will be predictable and the inevitable price rises that will happen in the coming years will be accompanied by matching wage increases.

Biochar - *by James Bruges*

Charcoal retains the carbon cell structure of plants from which it is made and, when buried, the carbon can stay in the ground for hundreds or thousands of years. Most fertile soil contains charcoal from ancient or recent forest fires and, until the introduction of synthetic fertilisers, charcoal was widely used by cultivators for enhancing the soil. The most remarkable example of soil modified by charcoal is the deep 'terra preta' from a previous civilisation in the Amazon that transformed infertile earth into rich loam.

'Biochar' is a new term applied to charcoal that is specifically produced for agricultural purposes. It differs from the familiar charcoal used for barbecues or cooking, which is made from wood in relatively large chunks and retains volatiles that increase flammability. Biochar, on the other hand, can be produced from any biological material and the producer should ensure that it is free of volatiles: wood vinegar, for example, is a valuable by-product as a pesticide and if retained in the biochar would hardly help microbial life! Volatiles and ash could also block the entrance to cavities and make them unavailable to moisture, microbes, fungi and mycorrhizae. Water-holding capacity, therefore, is one of the easiest tests for assessing the quality of a useful biochar because it indicates whether the cavities are open and available. Another indication is the proportion of char to feedstock: a high proportion may indicate that volatiles and ash are retained, adding weight and rendering the char of little, or no, use for soil enhancement. The temperature at which biochar is produced is critical for achieving a high degree of absorption. The cavities give refuge to microbial life that forms a bridge for nutrients to the hairs on plant roots, thus enhancing fertility. It is desirable to mix biochar, before use, with manure, digestate slurry, compost etc. for it to be 'charged' and have an immediate effect with plants.

One hope for biochar, therefore, is for increased global food production while permanently enhancing soil. The other hope is that it could help the struggle against climate change.

Carbon dioxide in the atmosphere is mobile. The 'carbon cycle' refers to the movement of carbon from sky to earth and back again. Plants capture carbon through photosynthesis then the activity of microbes release it back to the atmosphere; there is about three times as much carbon in soil and plants as in the atmosphere. Over a period of 14 years an equivalent of the entire volume of atmospheric carbon is captured by plants, enters the soil and is eventually released back to the sky. If more carbon can be stored in the soil there will be less in the sky. Human activity is adding a steady stream of carbon to the atmosphere. Oceans have the biggest store of carbon but excess carbonic acid is already a serious problem.

Land has lost carbon since the dawn of history through overuse or harmful farming methods. The Fertile Crescent is an obvious example. Synthetic fertilisers interrupt the carbon cycle and degrade soil by providing nutrients direct to plants. Compaction from heavy machinery is another problem, and rising temperatures are now reducing the ability of soil to retain carbon. It has been estimated that land in the UK may have lost about half its embedded carbon

content since the industrial revolution, so there is plenty of scope for increasing carbon in the soil without harming natural systems.

Organic and permaculture cultivation retains carbon in soil. Not much research has been done in temperate climates on further fertility benefits from biochar though, of course, it would also add carbon. Temperatures above 25degC lose more carbon to the air than in cooler climates, giving a greater tendency to desertification. In dry tropical areas the ability of biochar to retain moisture is the first aspect that appeals to farmers, while its ability to improve the structure of soil and provide a haven for microbial life is equally important. Biochar fertiliser is compatible with NPK fertilisers so can be used in a gradual process of reducing the latter in both temperate and tropical areas. In passing it is worth noting that some research indicates soils enriched with biochar curtail the emission of nitrous oxide.

Since the scientific study of biochar is relatively new there are widely differing views as to the part it might play in combating global warming. Much research has gone into some aspects of the technology but there are not many peer-reviewed papers relating to the performance of biochar in the soil at significant scales or over a number of years. This makes some scientists, concerned for their reputations, reluctant to include biochar as a major element in their recommendations for climate mitigation, whereas the contribution of trees is fully understood. However, the principles set out above indicate that biochar has a huge potential that should attract government funding unrelated to commercial incentive.

Most commentators on climate change talk about reducing emissions. However, global warming is caused by excessive greenhouse gas already in the atmosphere, so merely reducing emissions will not dig us out of the hole. Carbon dioxide must be extracted. Chris Goodall wrote *Ten Technologies to Save the Planet* about means to tackle climate change; eight are concerned with reducing emissions and only two are for extracting greenhouse gases from the atmosphere. Biochar is one of these. James Lovelock has said biochar in the hands of farmers around the world may be our only hope. Others suggest it may play a small role compared with prevention of deforestation and extension of tree planting. Then some commentators say that biochar is just one out of may geo-engineering proposals. It is not. It enhances natural processes whereas most others may have very dangerous unforeseen consequences.

Like all major new technologies there are some down sides. If the burial of biochar were to earn carbon credits there would be a commercial incentive to grow monocultures to provide feedstock for biofuels and biochar: taking the food out of hungry mouths to feed the cars of the rich. However carbon credits are unlikely to be approved because of difficulties over accreditation (how much has been put in the ground? how long will it remain? how much has been burnt? how efficient has the production process been? What is the carbon-effect of monocultures? etc.). There is a campaign that relates possible down sides like monocultures with the harmful effects of biofuels. It focuses on unknowns and any published statements that indicate biochar may not always be effective. In this it is amusing to see similarities in arguments used by deniers of climate change over the last decade: the former are on the traditional left and the latter on the right of opinion. However these strident voices may be useful in alerting governments to the need to have a regulatory framework and not leave the process to the outlandish distortions of the free market.

There are many scientific caveats to the above paragraphs but they give an idea of the importance of plants and soil when thinking of ways to increase food production and reduce excessive carbon concentrations in the atmosphere.

Albert Bates, The Biochar Solution, 2010.

BiofuelWatch, campaign group against biofuels and biochar.

James Bruges, The Biochar Debate, 2009.

Johannes Lehman & Stephen Joseph, Biochar Environmental Management, 2009

James Lovelock, Biochar as Solution to Global Climate Change, YouTube, 2009

Chapter Seven

Cap and Share: managing the share on a global level

Caroline Whyte

There are two key questions that arise when one considers the practicalities of distributing the share in Cap and Share – or any other per-capita-based share of a common resource – globally: whether it would be physically possible to distribute equal allocations to every single adult or every single person in the world, and whether it would be wise.

Let's being by running through some of the main objections that spring to mind when we consider how a global cap and share scheme might play out:

- **The instability of debt-backed money within the 'casino' world economy.** Relying on national currencies in their current debt-based form as the primary way to express the value of the share seems likely to prove unwise, to say the least. At the time of writing we're being reminded yet again of just how fragile the world's financial system is. By the time this book is published, it's entirely possible that the euro will have collapsed, bringing the world's stock markets down with it.
- **Lack of infrastructure.** This problem could affect the distribution of the share, since getting emissions allocations or cash to people who live off-the-grid and far from any road is a challenge, and it could also affect the extent to which the share is useful. There's not much point in having cash if there's nothing meaningful in the locality to spend it on.

- **Unstable regions.** Large areas of the world are either in a state of war or barely out of war, and the logistics of getting shares to refugees and displaced people could be daunting.
- **Possible gold rush effect.** Local economies could be destabilized by a sudden inflow of cash, as has happened before in areas where a valuable natural resource was discovered. In this case the resource would be emissions rights, and as with other resources, its 'discovery' could trigger problems such as inflation. It could be particularly damaging to communities which are not very dependent on cash at present. Moreover, as with other resource discoveries, an economic boom in the area would likely be ephemeral and might quickly give way to a bust, particularly if the world economy collapsed because of energy shortages.
- **Problems related to the temporary nature of the scheme.** If the cap is successful in bringing emissions down to zero over time, the scheme will end. But there could be resistance to the gradual winding down of Cap and Share both on the part of the people employed to carry it out and of those beneficiaries who gain the most from it.
- **Possibility of violent crime.** Some areas of the world are already in a state of crisis because of climate change or other disasters, and desperate people might take extreme measures to get hold of as much cash as they can. They might also spend it on weapons. Vulnerable groups such as women and the elderly could be targeted.
- **Vested interests.** As we'll see, one of the effects of allocating the share would likely be a decrease in inequality. This could be perceived as very threatening by people who benefit from the status quo, such as big landowners in Brazil. They tend to have a lot of political clout and could make it difficult to implement the programme.

This list may seem quite forbidding. However, the world situation is evolving very quickly and there are three recent phenomena which we can factor against the objections listed above. A decade or so ago, none of them would have been weighty enough to make much difference, but as we'll see, circumstances are different now. Let's take a look at each of them in turn.

Increased recognition of the importance of empowering individuals and local economies

Broadly speaking, it used to be widely assumed that developmental planning was something best undertaken either solely by governments or by a combination of governments and big businesses who would undertake massive projects such as dam construction[1]. Then, in the eighties and nineties, government went out of fashion and there was a trend towards privatizing publicly held resources such as water and energy. As Justin Kenrick and Nick

Bardsley point out in their chapters, agencies such as the IMF and World Bank have become notorious for pressuring countries to adopt these policies.

In some areas such practices continue to the present day. But it is increasingly recognized that there are enormous problems with putting all the big planning decisions in the control of large bodies, be they public or private. Such problems include corruption, a systematic, relentless transfer of wealth from the poor to the rich, too much bureaucracy, a lack of accountability and misperceptions as to what 'ordinary' people actually want and need.

An important aspect of the reaction against this is the global commons movement which Justin Kenrick describes in his chapter. I'll be discussing another aspect here: the increasing amount of attention that has been given over the past decade and a half to "social transfers".

Social transfers are anything useful that lends itself to per-capita distribution by governments or NGOs. They can take the form of cash, food or vouchers, among other things. Cash transfers in particular are becoming increasingly popular, despite initial fears on the part of some observers that they would create dependency and stifle initiative.

In their book *Just Give Money to the Poor: The Development Revolution from the South*, Joseph Hanlon, Armando Barrientos and David Hulme write that cash transfers are a "southern challenge to an aid and development industry built up over half a century in the belief that development and the eradication of poverty depended solely on what international agencies and consultants could do for the poor, while discounting what the citizens of developing countries, and the poor among them, could do for themselves".[2] They add that "the biggest problem for those below the poverty line is a basic lack of cash. Many people have so little money that they cannot afford small expenditures on better food, sending children to school, or searching for work." , and they cite statistics showing how cash transfers have substantially reduced poverty in countries as diverse as Brazil and Mongolia.

They note that poorer people who receive transfers are more likely to spend the money on locally produced goods than richer people, who tend to spend more money on imports. Additionally, poorer people are more likely to use the transfers as leverage for investments, rather than simply spending all the money immediately.

According to a recent report by the International Labour Organization (ILO) and the UNDP, "there is growing international consensus on the importance of essential social transfers and essential social services as core elements of a social protection floor for national development processes." Many charities and other NGOs are now strongly promoting social transfers. Even the World Bank, which is not noted for its radical views, has come round to endorsing them, albeit in a form which some argue to be unnecessarily complex[3]. Indeed, I have been unable to find a single report or study that considered social transfers to be a bad idea, although, as one might expect, some individual programmes come in for criticism because of being poorly designed or implemented[4].

Social transfers can be short-term, with the goal, for example, of providing emergency aid to people who are having difficulties, or longer term, to help provide financial stability so that people can make investments and plan realistically. Sometimes they come with strings attached, as in various Latin American countries where people are guaranteed a stipend if they enroll their children in school. These are known as conditional transfers. They can be targeted, i.e., given only to people who fall within a certain income range or live in a particular area, or they can be universal.

So how would Cap and Share fit in with this? It seems clear that the share in Cap and Share could be considered to be a universal, temporary, unconditional social transfer, to be allocated by an NGO which would probably take the form of a trust. We should note however that there are three differences between it and existing social transfer programmes.

The first difference, which is probably not very important, is that the goal of existing social transfer programmes is usually either poverty relief or 'development', purely and simply; such programmes generally make no reference to the environment or even to commons-based rights. This difference in purpose would have no effect on the mechanics of distributing the transfers, and in any case, if Cap and Share were to be carefully implemented, it would be quite likely to substantially reduce poverty, at least in the short term (see panel).

The second difference is in the source of the transfer funds. Existing social transfer programmes are all funded from taxes or from donations to charities, and so take the form of wealth redistribution. The share in Cap and Share, however, is a form of *pre*distribution since it derives from the natural commons of the atmosphere. Thus, it might not be subject to some of the short-term political pressures that tax-funded social transfer schemes can sometimes fall victim to, which would of course be an advantage. Indeed, Hanlon *et al* actually suggest towards the end of their book that the funding from carbon use fees be used internationally as cash transfer revenue; their rationale for this is that funding for the transfers would be more reliable under such a scheme than tax- or loan-based funding. So these authors have independently developed an idea that comes quite close to Cap and Share.

Cap and Share in India and South Africa

Two reports commissioned by Feasta in 2008 examined the possible effects of a global cap and share scheme. The first one discussed the South African situation and the second focused on India. The South Africa report[5] was written by Jeremy Wakeford of the South African New Economics Network. Here is the section of its conclusion that describes the effects of the share:

> *"South Africa mirrors the wider world in that it has two linked economies, a rich, energy-intensive one and a poor, low-energy-use one. As a result, this study of the effects that Cap and Share would have if introduced there as part of a global climate settlement provides a good indication of how C&S would affect the world as a whole. What it shows is that:*
>
> - *70% of the population would be better off because they would receive more from selling their emissions permits than their cost of living would go up. The income of the bottom 20% could double.*
> - *The richest 20% of the population could see its income reduced in the short-term by 14% while the 10% of people with middling incomes would be unaffected because their increased costs would be balanced by their increased income from selling their permits.*
> - *The higher fossil fuel prices C&S would bring would provide the incentive for a rapid development of renewable energy sources."*

India still has a comparatively low level of per-capita emissions, since even though it has a booming economy and its emissions levels are increasing rapidly, these factors are counterbalanced by the fact that emissions started at a relatively low level and the population is large. The India report, by Anandi Sharan of the Bangalore-based NGO Women for Sustainable Development, therefore concluded:

"Because most of its people use very little fossil energy, India would benefit massively from the global adoption of Cap and Share [...] 90% of the population would stand to gain, and the more rapidly emissions were reduced, the greater their gains would be. For example, if the pace was rapid and the world price for emissions permits rose to €100 per tonne of CO_2, the poorest 10% of Indians would see their total income increased twenty times. If the CO_2 price was €200 per tonne, their incomes would be 40 times greater than today. Only those Indians using a lot of energy would suffer in the short-term because they would have to pay more for their fuel than they received in compensation from the sale of their permits. Their net incomes would be reduced by 0.26% at €100/tonne of CO_2 and 0.52% at €200. In the longer term, however, they could expect to become richer as a group because of the better business and professional opportunities the increase in the rest of the population's incomes would provide." [6]

We can see that in both countries there would be a huge reduction in income inequality. The sudden rise in income of the lower-income population deciles would be likely to trigger radical shifts in the countries' economies, and indeed in their societies and cultures, and the power relations between these countries and the rest of the world would also change. There is plenty to think about here, and we'll discuss some of the ramifications further below.

The temporary nature of the share

The third difference is likely to be the most important one: it is that the share in Cap and Share would not be a dependable, fixed amount of money, but would fluctuate in value. At the beginning it would quite possibly be far more valuable to most of its recipients than any social transfer made at present (with the exception, some would argue, of financial bailouts made to billionaires on Wall Street – a rather different kind of social transfer from that promoted by the UNDP and ILO). As mentioned above, in some areas its value could be so great that it could even destabilize the local economy. Later on, its value would depend on the extent to which the world economy was succeeding in transferring to renewable forms of energy. This places it in sharp contrast with current social transfer programmes, which only deal with modest amounts of money that do not fluctuate.

What are the implications? Obviously, Cap and Share could not be part of a "social protection floor" as it would not provide a reliable income or food security. However, the emphasis placed by social transfer schemes on providing individuals with decision-making power about how to allocate resources dovetails very neatly with the commons- and resource-based philosophy behind Cap and Share. Many charities express their support for social transfers in terms of individual rights and agency.

Moreover, if Hanlon *et al* are right, any distribution of wealth in favour of the poor will tend to stimulate local economies, a fact that has important ramifications for cutting down on fossil fuel use: it means that the share in Cap and Share could reinforce the cap by creating an environment in which locally-based industries are able to develop and flourish. It could also be pooled by community members and used to secure customary land tenure systems such as Justin Kenrick describes in his chapter. In addition, the investment possibilities for people receiving funds would certainly apply to people looking for uses for the share in Cap and Share. The money could be treated as a nest egg to be used for projects that would improve their lives, such as education or better housing.

It would clearly be important to make sure that beneficiaries were getting accurate information about Cap and Share in order to avoid confusion. If it was understood that the scheme was temporary, they would be likely to make very different decisions about what to do with their allocations than if they were under the impression it was permanent.

However, we should note that even though Cap and Share itself would be temporary and the value of the allocations would fluctuate, there is no reason why it couldn't serve as a springboard for a more stable permanent transfer scheme, such as a basic income derived from a Tobin tax or land value tax. The same databases and communications networks could be used for such long-term schemes, and Cap and Share could perform a useful role in providing the initial capital to get the overall system going while the details of establishing a source of revenue for a more permanent scheme were being hammered out. This might have the convenient side-effect of warding off some of the potential problems with Cap and Share being a temporary scheme: employees would have more security and beneficiaries would know that they could count on receiving transfers in the longer term[7].

Techniques for the distribution of emissions allocations

Let's move on now to examine the second global trend which could provide a boost to a Cap and Share: the widespread adoption of technologies which could be used for distributing the share. If we look at the many experiments that have been made with distributing social transfers we get some idea of the possibilities. Those which have the most potential for overcoming distribution challenges are probably smart cards and mobile phones.

Namibia, Botswana and South Africa now use **smart cards** for their universal pension schemes. Once the smart cards have been created there is no need for any other identification, since smart cards can record people's fingerprints, which can then be checked with a simple digital camera each time a transaction is made. Recipients can opt to keep some or all of their funds in the card account. They access the funds through banks, mobile ATMs or through point-of-sale terminals in shops.

The East African Regional Hunger and Poverty Programme comments that "even street-traders and village merchants are now clubbing together to share the use of such low-cost terminals."[8] This means that recipients of the pension do not need to live anywhere near a bank or post office in order to make use of their allocations – they need only be in striking distance of a market with vendors who use the terminals. They could opt to withdraw only some cash or to go entirely electronic when paying the merchants with the terminals. Such terminals can be run on solar power[9], so areas which are not on the electric grid could be included in the scheme.

Mobile phones are only just beginning to be used for social transfer allocation, but they are also showing considerable potential. In a pilot scheme in 2009 jointly organised by the Kenyan government, the charity Concern International and the mobile phone company Safaricom, beneficiaries in a very remote rural area of Kenya, off the electric grid, were each given a SIM card. Mobile phones were also distributed among the population, one for every ten households, together with solar chargers and handsets. Each time a transfer was due, an SMS message would be sent to the mobiles explaining where to go to cash in the transfers, and the beneficiaries would walk there (a maximum of 8 kilometres) and collect the money from a mobile ATM. Obviously someone in the ten households had to be literate enough to read the SMS message, but in practice this was not a problem. The cost of the scheme was calculated at 5%, which is low compared to that of many earlier transfer schemes.[10]

As with smart cards, beneficiaries could spend some of their allocation immediately on food or other provisions if they wished, and they could also opt to keep some in the mobile account for later. This ability to manage money wirelessly and store it securely by electronic means is so convenient that a New Statesman article from 2008 argues that mobile phones are likely to become dominant as the way to handle money in the future.[11] (Previously, people had to send wire remittances if they wanted to transfer money from place to place, which were cumbersome and costly.)

Could mobile phones help to overcome the challenge of distributing shares from emissions allocations in areas of the world with little or no industrial infrastructure?

Photo source: http://www.sxc.hu/photo/1125851 Author: m_a_essam

A final point we should note about mobile phones is that in addition to being used for financial transactions, they are, as one might expect, still doing their original job – putting people in contact with each other, or to put it more trendily, building social networks. As of late 2008, there were almost 4 billion mobile phones being used worldwide, with the bulk of the market growth taking place in the BRIC countries[12][13]. This opens up some interesting possibilities for Cap and Share, which we will explore further below.

However, we should first consider another question that was mentioned at the beginning of this section with regard to allocation distribution: the potential for violent crime. This could occur in the very direct form of gangsters who might wish to steal allocation funds. More generally, there is also the challenge of allocating emissions rights in the many parts of the world that are highly unstable. How could a social transfer programme possibly function in such situations?

In a 2006 book on designing social transfers, Michael Samson of the South African Economic Policy Research Institute asserts that "Even in fragile states – such as Nepal and early-1990s Mozambique – governments have effectively delivered social transfers." He describes a transfer scheme that was implemented in Mozambique in the 1990s, during the civil war there, and comments that "the programme worked remarkably well in the first five years"[14] (later on it foundered due to administrative and funding problems, and it has since been replaced by another scheme).

Various techniques can be used to ward off would-be robbers. For example, in areas where a mobile ATM machine might be vulnerable to hijackers, agencies use tactics such as changing the route that the vehicle carrying the ATM takes each time it enters the area, and having it stop to distribute cash in different places as well.[15] In fact, I came across only one direct reference to a hijacking of a vehicle carrying cash intended for social transfers, in Uganda, and this hijack was considered rather exceptional – it was an inside job, carried out by people working on short-term contracts for the agency. The report which described the hijacking concluded that longer-term transfer schemes in which the distribution work is contracted out are much less vulnerable to this kind of problem: "Using [such a] system goes a long way towards minimising the possibility of insider mischief; agents are not short-term contract staff, they are entrepreneurs looking to build a sustainable business so have a vested interest in the system working safely [16]".

One might wonder, though, whether the large amounts of money likely to be involved with Cap and Share, at least initially, might not create a greater incentive for violent crime. After all, most existing cash transfers involve only modest sums, not enough for it to be worth a gangster's while to chase down recipients after they have collected their cash and force them to hand it over. With Cap and Share we could be talking about serious money, as is suggested in the reports on India and South Africa.

One way to deal with this could be to encourage people to use their smart card or mobile phone accounts as safe storage places for their money, and only withdraw cash in small amounts as needed. There may also be other good reasons for people not converting all of their emissions allocation into the local currency for immediate personal use, as we shall see.

However, it could be that the threat from crime is somewhat overblown, particularly if we take into account the ramifications of the share. As mentioned above, one of its effects would likely be a decrease in income inequality – indeed, most social transfer programmes have this effect [17] – and there is a great deal of evidence that as income inequality decreases in a country, so too does violent crime [18.] Of course it should not be discounted altogether, but it may be less of a problem than one would fear.

One other point we need to discuss when considering the use of modern technologies is the possibility – even probability – of system collapse. Some analysts believe that the world's communication infrastructure is actually very fragile – it is, after all, intermeshed with the wider growth- and fossil-fuel-based economy – and that a trigger such as a shortage of oil could send it over the edge, causing the entire communications network to rapidly unravel[19]. In such a case, obviously it would be impossible to continue using it to allocate the share.

However, the fact remains that the communications network is the most efficient means we have to allocate the share at present, and it makes sense to use whatever is available to get things moving quickly. Even if our success in allocating the share is short-lived because of a system collapse, there's a reasonable chance that we will have improved the medium-term and perhaps even long-term stability of many regions of the world by decreasing inequality, helping to secure key commons rights and encouraging investment in locally-based energy and food production. This will help many ordinary people cope with the coming shift to an economy without fossil fuels.

Moreover, there's a certain poetic justice in using the very technologies that are considered by many to symbolize an exploitative and rapacious economic system to help to bring that system to as smooth an end as possible. Or, to put it less loftily, simple decency would suggest that we put these things to as good a use as we can while they're still available to us.

Effects of share distribution

We have taken a look at some techniques for distributing allocations, and this hopefully has allayed at least some of our fears that such a distribution scheme would be fundamentally unworkable. Now let's turn back to some of the other challenges described earlier: the lack of good things for individuals in some places to spend share money on, and the possibility that the large influx of money that could take place in some areas would cause inflation or destabilize the economy in other ways.

With regard to the latter problem, as already mentioned, if we look at past experience we can find many examples of per-capita distribution of resources, but thus far the amount of money distributed to each person has been relatively modest. We can also find plenty of examples of areas which experienced a sudden, large inflow of funds, such as gold rushes. However, to the best of my knowledge there has never been an occurrence of both together – or in other words, a situation where every single adult member of a community has received a series of rather large sums of money. The closest thing is probably the Alaska Permanent Fund, which allocates a part of its revenue to every adult citizen of Alaska each year. This is around $2000 per person at the moment[20] and for some people it represents 10% of their annual income, but we have seen that Cap and Share in India could dwarf that, increasing the income of many people fivefold or more.

Historically, gold rushes have been characterised by their unpredictability – the fact that nobody knew in advance that they would take place – and by the unevenness with which the revenue from the newly-discovered resource was distributed, with some people becoming wildly rich in a very short time while others were left behind entirely. Both of these factors have a highly destabilizing effect on an economy. However, neither of them would apply to the share in Cap and Share: everyone would know in advance that they would be receiving money, and everyone would get the same allocation.

Even so, it seems clear that we shouldn't dismiss the possibility of destabilization. For one thing, countries which experience a sudden surge in income because of the discovery of an important raw material can sometimes run into problems with exports because their currency gains a lot of value owing to demand for that raw material, making their exports more expensive ('Dutch disease'). In the case of Cap and Share the "raw material" would be emissions rights, and countries whose inhabitants mostly didn't use much fossil fuel, such as India and much of Sub-Saharan Africa, would be "exporting" the rights elsewhere. Their currencies might then rise in value, which could adversely affect other sectors of their economies; many of these countries export other raw materials, and it's possible that they would experience a slowdown in those exports. But on the other hand, the increase in value of their currencies would make it rather easier for them to pay off international debts, which have had a crippling effect on their economies[21]. Indeed, this ability to pay off debt painlessly is a strong point in favour of Cap and Share.

As with any macroeconomic issue, there are enormous complexities and unknowns here. We could probably get a clearer idea of the effects of the share in those countries where it would be very valuable to individuals by doing a more thorough study of the historic effects of gold rushes and other "rushes", while taking into account factors such as the degree of income inequality already existing in places that experienced them, and the extent to which the incoming wealth was distributed among the population.

Another possible problem we've mentioned above is that the sudden increase of cash in local economies would lead to inflation of prices for basic staple products and perhaps also for land. People would still need the same essential goods that they needed before the money arrived, but there would be a lot more money chasing the same goods – a classic recipe for inflation. They would also be looking for secure places to park excess money and so there could be a scramble to buy up land. Since land is always limited in supply, there could then be all kinds of problems with some people being left behind in the dust while others in their communities charge up the property ladder, and with speculative bubbles forming. In the long term, a land value tax such as Nick Bardsley describes would take care of such problems, but very few places in the world have implemented such a tax at the time of writing.

With a view to these threats, let's assume that there would be a strongly destabilizing effect if we simply divided up the allocations and made them immediately tradable into cash for everyone, and look at a way to try and forestall it.

French taxpayers subsidise theatregoers

The problems described above could easily be assumed to derive from some fatal element of human nature – the greedy, grasping side to human beings. We're all frail and flawed, after all.

Indeed, some may argue that it's naïve to believe that ordinary people can be trusted at all to make sensible decisions about 'windfall' cash that they receive. Perhaps it would all be spent on beer or fast motorcycles.[22]

While I can certainly understand – and share – a certain skepticism about the benefits of money in itself, we also need to be careful not to make assumptions about what 'ordinary' people are capable of which could turn out to be rather patronizing. As mentioned above, experience with existing cash transfers has shown that they tend to be used carefully and sensibly. So there are practical reasons to take the per-capita approach. But there are also strong reasons having to do with ethics and justice: if the 'ordinary' people aren't deciding where the allocation money should go, who is? What right do they have to do so?

And in any case it seems much more accurate to regard the source of this particular problem as deriving from the nature of money rather than that of people. Specifically, it's the assumption that the share would have to be in frail, flawed, debt-backed, bank-issued, bond-market-dependent money that could lead to problems. Here we're led back to the very first problem from our list at the beginning of the chapter: the knack that our money has for becoming 'funny', in an unamusing kind of way.

So perhaps we should take a creative approach to the type of money that is used for the share.

In France, if you have a young child or are retired, you are entitled to a certain amount of "cheques vacances" per year. These can be spent like cash in a wide variety of places, such as campgrounds, hotels, restaurants, amusement parks, and theatres. Businesses can apply to accept these cheques – they just have to be involved in some way with recreation in order to qualify. The cheques expire after three years, so once you get them you have an incentive to spend them.

One effect of this programme is that individuals who might not have found the time or energy to go on holiday are much more likely to do so. It also ensures that the economy as whole is stimulated in a way that it might otherwise not have been. Those businesses who accept the cheques will benefit financially, and there is a multiplier effect in their local economies.

So perhaps a helpful approach would be to make the share in Cap and Share tradable with something analogous to cheques vacances, rather than normal currency. Let's call them clear air cheques. In a Cap and Dividend-type scheme, for example, the trust would auction off emissions permits and collect the revenue in ordinary money. Then it would issue an equivalent amount of clear air cheques to the population on a per-capita basis. (They wouldn't have to be paper cheques of course – they could simply be credited to "clear air accounts" that people would have on their mobile phones or smart cards).

The clear air cheques would have an expiry date, so they wouldn't be hoarded, and they could be spent just like cash, but only on certain things. These could include legal aid for the securing of commons rights, renewable energy projects, and investment projects that would not use a great deal of fossil fuel – ideally, none at all – but that would nonetheless be important to community wellbeing, such as health care and education, or that would support carbon sequestration activities, such as organic farming and tree planting for agro forestry and water catchment protection. Those receiving the cheques would be able to redeem them with the trust and get regular money back[23].

For the recipients of the allocations, this would be a very similar set-up to the pay-as-you-go accounts that many mobile phone users have, which also often expire after a certain length of time if they are not topped up. In fact, it should be no harder to grasp than one of those accounts is: the allocation would go in in much the same way as a top-up and then you would use the credits. For the retailers, it would be like dealing with any voucher or coupon, or a currency different from the usual one they deal with – it would involve slightly more work than regular cash but would be worth it for the extra business. All the transactions could be handled electronically with no more fuss than existing mobile phone accounts.

Some readers may now be thinking that this is all very well, but how would it help with the lack of infrastructure? As we already noted above, there is no use in having a lot of money available for individuals to pay for their childrens' education if all the schools in an area are overcrowded and badly designed, or nonexistent. Health care facilities are also very thin on the ground in some areas of the world, and renewable energy technology, which would be an ideal thing for individuals or small groups to spend Cap and Share money on, is probably still in its infancy.

One approach might be for the trust that administers Cap and Share to divide up the revenue from the permits and use some of it for top-down infrastructure development. It may well be a good idea to use part of the revenue set up a Children's Fund as suggested by Laurence Matthews in his chapter. However, the greater the portion of the revenue that was used in this way, the further away we would arguably get from the idea of treating the atmosphere as a commons, as it would take some power away from ordinary people who would no longer have their total share of the revenue. It takes us further away from the principle of subsidiarity which John Jopling describes in his chapter, that suggests that decisions should be made at the lowest possible level. We could then run into the same old problems with bureaucracy, corruption and inflexibility that arose in the era when top-down developmental decisions were considered the only way forward, not to speak of the ethical issues described above.

In any case, much of the development needed is actually on a small or intermediate scale and decentralised, rather than on a large scale and centralised – things like solar power installations, primary health care clinics and schools which function best when they are not too large and unwieldy -in addition to organic farming methods and agro-forestry which would address another aspect of commons erosion. Such investments make most sense when done on a community level, rather than by individuals acting entirely separately or by regional or state governments.

Here is where we get to the interesting part. Since mobile phones are so helpful for social networking, they could possibly be used to help make community-wide decisions about how to allocate larger amounts of money. Whenever members of a community heard that allocations of the share were on their way, they could hold a meeting and compile a list of projects that they considered to be a priority for the whole community, such as getting legal aid for establishing commons rights, building a school or a primary clinic, or sending one or two people from a village to a Barefoot College to train as solar engineers. People could then pool their clear air to pay for the projects. They would have choices as to which projects they wished to prioritize and they could "vote" with their mobiles or smart cards to allocate their funds accordingly. In this way the decision-making process would be kept as broad as possible (and out of the hands of local élites who might otherwise co-opt it), but the local currency would be spared the burden of dealing with everything at once: much of the money would be kept out of immediate circulation for dealing with larger, medium-term or long-term community-based projects[24].

Guatemalan schoolchildren. Individuals could pool part of their share to build infrastructure in their community for schools and hire teachers.

Photo source: http://www.sxc.hu/photo/253109 Author: jwarletta

Obviously this is just a rough overview and there are many ways in which a system such as this could be fine-tuned. For one thing, it would probably be a good idea to make at least a small portion of the allocation convertible into the local currency right away, rather than into clear air, so that people with immediate needs for essentials such as food could meet them[25].

For another, as Laurence Matthews mentioned in his chapter, there are some areas of the world that are not in the cash economy at all, and people in those areas might well prefer to keep them that way. As mentioned above, many cultures and traditions believe that money should be treated with extreme wariness, and for good reason. Individuals or communities should therefore be able to simply opt out of the scheme if they wish, or to donate some or all of the proceeds from their allocations elsewhere [26]. This decision needs to be made by 'ordinary' people, though – not by powerful people acting on their behalf, including powerful people from within their own country or culture. Again, the logistics of this is something that communication technologies could help with.

One other point to make about the share is that it represents an entitlement – a right – rather than justice in and of itself. What I mean by this is that per-capita shares follow the same logic as one-person, one-vote political systems. Just as voting rights grant individuals a certain amount of power but we also need a legal system to handle issues of justice, the share in cap and share ought to represent a step towards greater justice for those who are disempowered by the current world situation, but certainly shouldn't be considered to deliver complete justice. Much more is needed for that, possibly including the legal enforcement of Greenhouse Development Rights.

In the meantime, while such legal issues are being thrashed out, we could at least get on with distributing the share. We shouldn't lose sight of the fact that it could save many lives in the short term.

But what about the rich?

Observant readers will have realised that there's still a problem, though. It might be possible to distribute the share effectively, and it might prove very useful in the effort to reestablish commons-based land use and to eradicate poverty, as well as for encouraging investment in a renewable-energy-based economy, but there is a powerful élite in the world – the top 30% – which may not be terribly happy to see those things happening, particularly if they believe that it would impact on their own wellbeing.

This brings us onto the topic of power dynamics and the ways in which the less-powerful relate to the more-powerful in the world. That's clearly an enormous subject. I'll limit the discussion here to possible triggers for decreases in income inequality, since that is something that relates directly to Cap and Share. What sorts of things make the rich willing to fork over some of their money to the poor?

It's easy to get the impression from the media that the gap between rich and poor is widening all over the world right now; that's certainly the case in some important countries such as China and India. However, there are also exceptions to this trend, and if we look at real-life places with decreasing inequality we can see that a certain political momentum can build which makes it quite difficult to undo various changes once they have taken root. The clearest examples of this are probably Brazil and Mexico, both of which have notoriously powerful and unscrupulous elites, but which also have extensive social transfer programmes.

Mexico remains a highly unequal society, but over the past two decades its level of inequality has begun to decrease. A UNDP study[27] on the reasons for this concluded that it is caused by a better-educated workforce and by its massive social transfer programme. In fact, there's a connection between these two things because the social transfers encourage people to enroll their children in school.

A quarter of Mexico's population, or 5 million households, is covered by its Oportunidades programme, which started in 1998 and has been greatly expanded over the years. Under the programme, families receive cash transfers in exchange for enrolling children in education programmes and ensuring they have regular health care check-ups.

What's particularly relevant to our argument is that the programme has proved to be politically robust. According to the UNDP study, "The programme is [...] notable in having survived not only a change of administration (no other major anti-poverty initiative over the past two decades has done this), but also in having survived the first change in 70 years of the political party in power. In fact, rather than discard the programme, the new party's administration changed its name from Progresa to Oportunidades, and starting in 2001 the new government increased coverage from 2.3 to 4.2 million households (mainly in rural areas), and added semi-urban and urban localities to the already established rural ones." [28] During the election of 2006, the two biggest parties both put considerable effort into claiming credit for the programme.

We mustn't be too starry-eyed about this; an article in Policy Studies from 2009 claims that "both Progresa and Oportunidades were specifically aimed at appeasing and depoliticising the increasing presence of resurgent popular movements, whilst acting as a bromide for the masses, so as to signal political stability and a disciplined labor force"[29]. There seems little doubt that Oportunidades played a major role in enabling the re-election of the right-wing PAN party in 2006, which is hardly a champion of the poor. At the time when Oportunidades was extended, corresponding cuts were made to general social welfare programmes.

However, we also mustn't ignore the fact that Oportunidades has genuinely improved life for a great many people, not only by increasing their immediate income but also by improving their health – including life-and-death health issues such as infant mortality rates – and their nutrition and education. If these people voted for the PAN because they perceived it as being responsible for such major life improvements, succeeding where others had failed, it's hard to blame them.

And if the author of the article is correct, then Oportunidades was implemented in direct response to popular movements. This doesn't mean, of course, that recipients aren't entitled to any more than they receive at present, or that they should allow themselves to become 'depoliticized'. What it does show is that there's a dialectic at work: the élite were responding to pressure from below. So what would happen if there was a push from below for a specific, easily-grasped, efficient scheme such as Cap and Share in a place like Mexico?

Now let's take a look at Brazil. As in Mexico, Brazil's major social transfer programme is popular and politically robust, and has achieved real results in reducing inequality. A New York Times commentary on Bolsa Familia, the Brazilian equivalent of Oportunidades, states that "today [.....] Brazil's level of economic inequality is dropping at a faster rate than that of almost any other country. Between 2003 and 2009, the income of poor Brazilians has grown seven times as much as the income of rich Brazilians. Poverty has fallen during that time from 22 percent of the population to 7 percent."[30] This is in a country that used to be among the most unequal in the world.

Also as in Mexico, the programme is a source of political wrangling, not over whether it should exist, but rather over who should be rewarded for creating it. A BBC article about it claims that "there seems to be a fairly broad consensus in support of such initiatives, with the only argument about who should get the credit for projects, some of which have been in existence for many years[...]The government of President Lula insists the investment is much bigger now, and that in recent years 20 million Brazilians have been lifted out of poverty." [31]

Bolsa Familia is popular "among all demographic groups", according to a 2010 Pew Foundation survey[32]. This doesn't mean that it's universally popular of course – a glance through the comments section of any online article about it will quickly demonstrate that it isn't – but it's worth exploring what might be making it attractive to at least some of the rich in Brazil.

In this case it seems unlikely that they view it solely as a political tool for getting their favoured parties elected. If that were so, it wouldn't be very effective, since the centre-left has been in power for some time now in Brazil. Perhaps there are other factors involved. What could they be?

Equal is beautiful

Here we come to the third phenomenon which could act as a boost to Cap and Share. This time the phenomenon isn't a trend or a technology, but rather a book that has already been mentioned in several other chapters: *The Spirit Level – why equality is better for everyone* by Richard Wilkinson and Kate Pickett.

Although many of us have always had a hunch that people were better off in more equal societies, it's been difficult up until recently to muster a coherent practical argument against the fear-based stance of those who believe otherwise. What could you say to argue against people who believe that human nature divides us inexorably into predators or prey, so that the only meaningful thing you can do in life is to look after your own, by the most ruthless means if necessary? The Golden Rule and other moral arguments in favour of fairness sound woolly and flaky to them.

I referred above to a major World Bank study which tracks the relationship between inequality and violent crime and shows that there is a strong correlation between those two phenomena[33]. What's particularly interesting, though, is that everyone in society is affected, not only those who are less-privileged. In a more equal society, the rich are also less likely to suffer the effects of violent crime.

The Spirit Level picks up on this study and on a great many others from the voluminous research carried out on equality over the last few decades. It demonstrates that the rich would benefit in almost every way imaginable from a more equal society. They would be healthier, longer-lived, better educated and less likely to become addicts. More equal societies also tend to be more innovative and their members are more willing to trust each other, qualities which could prove very helpful during the coming adjustment to a zero-fossil-fuel economy. This correlation between greater equality and wellbeing holds in places with widely differing cultures and levels of overall wealth[34].

In fact, the sole way in which the rich might not benefit from greater equality is that they might possibly have less money than they do now. And of course money is only valuable if you can spend it on things that you value. If people are willing to spend $230,000 on a guard dog to protect their property[35], it's possible that they might begin to see the sense of sharing out some of their money in order to help prevent their fellow citizens from turning to violent crime.

Of course, it's also possible that some of them are just too out of touch with reality to be reasoned with at all. It would be naïve beyond absurdity to assume that everyone in the top 30% will accept these ideas without demur, saying "ah yes, I see we were wrong all along to try and preserve our wealth at the expense of everyone else – here, take some of my money!" The book *Treasure Islands[36]* gives plenty of examples of people who are unlikely to be swayed by any appeal to share their wealth, even when the sharing-out would clearly work to their own benefit; their mental state is simply too pathological for them to be reachable.

The fact remains, though, that not everyone in the top 30% always behaves that way, as we can see from the attitude of rich Brazilians. Of course we don't know for sure whether the support expressed by some rich Brazilians towards Bolsa Familia derives from a feeling that life is better in a more equal society; it could also derive from a (hitherto unguessed-at) sense of basic fairness, or else perhaps from some deeply cynical line of reasoning that I can't fathom.

But in any case, the examples of Brazil and Mexico show that the political dynamics leading to social change are complex and not always what one might assume; the rich do not simply crush the poor, always and everywhere. While we certainly shouldn't blithely assume that everything will work out for the best, we should note that pressure from below can sometimes trigger real change, as seems to be happening in Mexico. And other factors may be at work too: it's possible that at least some rich people recognize on some level that they themselves would benefit in various ways from greater equality. So it's quite handy to have the hard facts regarding equality and wellbeing at our disposal while promoting a scheme that would increase equality. Sometimes facts do work against fear.

Possible Ways Forward

So to summarize: We've taken a look at two questions to do with global cap and share, namely, whether it would be possible to distribute the share globally on a per-capita basis, and whether it would be wise. We've noted how development theory is now broadly in agreement with the commons-based philosophy behind Cap and Share, which maintains that the decision-making process about resource allocation should be kept as broad as possible. We've seen how recent technologies such as smart cards and mobile phones could help to make the share distribution feasible, even in areas that are off-the-grid, or have high levels of illiteracy or crime. We've explored a possible way to use the new technologies in order to ensure that an adequate amount of investment is put into infrastructure development, while at the same time keeping inflation and other sources of economic instability at bay. And finally, we've taken a look at the political dynamics that underlie the implementation of existing large-scale allocation schemes, in light of the vast amount of research that has been done into the relationship between equality and wellbeing.

Endnotes

1. Some notable thinkers, such as Gandhi, challenged this approach to development from early on.
2. Hanlon, Joseph *et al* (2010), *Just Give Money to the Poor: The Development Revolution from the South*, Kumarian Press, page 4.
3. See for example "Catching People Before they Fall: Social Safety Nets Take Center Stage", World Bank website, http://web.worldbank.org/WBSITE/EXTERNAL/NEWS/0,,contentMDK:22632052~pagePK:642 57043~piPK:437376~theSitePK:4607,00.html. The World Bank favours conditional transfers, which is a matter of some controversy.
4. Many such problems with implementation, as well as bureaucratic delays, are caused by difficulties in figuring out who to include in targeted or conditional transfer schemes. Cap and Share would not be affected by such problems as it would be universal and unconditional.
5. Wakeford, Jeremy (2008). *Potential Impacts of a Global Cap and Share Scheme on South Africa*, Feasta, page 28
6. Sharan, Anandi (2008), *Potential Impacts of a Global Cap and Share Scheme on India*, Feasta, page 28
7. Such long-term transfers would probably be much more modest in size however, and so once again, accurate information would be important to beneficiaries..
8. "Delivering Social Transfers" (no date), Regional Hunger and Vulnerability Program. http://www.wahenga.net/sites/default/files/briefs/Brief_5.pdf
9. See for example the company Hypercom's line of contactless smart cards and terminals, described on their website here: http://www.contactlessnews.com/2010/05/10/hypercom-luanches-l5000-line-of-payment-terminals
10. Brewin, Mike (2008), *Evaluation of Concern Kenya's Kerio Valley Cash Transfer Project* April-June 2008. http://www.mobileactive.org/files/KenyaCashTansferPilot-EvaluationReport-July08.pdf http://www.concern.net/sites/concern.net/files/resource/2008/08/1193-kenyacashtansferpilot-evaluationreport-july08.pdf
11. "Hanging on a Telephone", Jack Hancox, New Statesman, 18 September 2008 http://www.newstatesman.com/society/2008/09/mobile-networks-families
12. "Half the World's Population Will Be Using Mobile Phone By End Of Year", *The Guardian*, 26th September 2008. http://www.guardian.co.uk/technology/2008/sep/26/mobilephones.unitednations.
13. For a fascinating glimpse into the ways in which mobiles are being used worldwide, see the textually.org website, and in particular their section on the developing world.
14. Samson, Michael *et al* (2006), *Designing and Implementing Social Transfer Programmes*, EPRI, page 15
15. ibid, page 129
16. Brewin, Mike (2008), *Evaluation of Concern Kenya's Kerio Valley Cash Transfer Project* April-June 2008, p 22. http://www.mobileactive.org/files/KenyaCashTansferPilot-EvaluationReport-July08.pdf
17. Prasad, Naren (2008), "Policies for Redistribution: The uses of taxes and social transfers", International Institute for Labour Studies, http://www.ilo.org/public/english/bureau/inst/publications/discussion/dp19408.pdf

18. Fajnzylber, Pablo *et al* (2002), "Inequality and Violent Crime", *Journal of Law and Economics*, vol. XLV (April 2002) http://siteresources.worldbank.org/DEC/Resources/Crime&Inequality.pdf. This study tracked the relationship between income inequality and violent crime in 36 countries over the course of several decades.
19. See for example David Korowicz's article 'On the Cusp of Collapse' in *Fleeing Vesuvius: overcoming the risks of environmental and economic collapse*, 2010 Feasta.
20. Source: http://www.apfc.org/home/Content/home/index.cfm
21. It has been suggested that a new world currency, the emissions-backed-currency-unit or ebcu. be introduced along with Cap and Share. See http://www.feasta.org/documents/energy/Cap-and-Share-May08.pdf. Ebcus would be allocated to each country according to their population, with the stipulation that a part of the allocation would have to be used to pay off international debts. This would be easy for highly-indebted countries to do as their currencies would be strong in relation to ebcus. However, it may not be wise to introduce a currency that is backed by something – carbon dioxide emissions – which we eventually want to eliminate altogether.
22. The beer-and-fast-motorcycles argument is very interesting because it touches on an important aspect of the world economy: the fact that it triggers stress which can then lead to unhealthy addictions. 'Shopaholism' and over-consumption in general have been linked convincingly by researchers to other, 'harder' forms of addictions with common roots in brain chemistry that has been distorted by a stressful environment. If the economy were less chaotic and more equitable, the stresses that drive people to spend money unwisely would probably diminish. See http://www.feasta.org/2011/07/16/is-over-consumption-hard-wired-into-our-genes/.
23. Of course, this "regular" money needs to be redesigned as its existence depends on the use of fossil fuels, as described elsewhere in this book.
24. Such community meetings could make use of already-existing political structures if these seemed appropriate , such as the Indian *panchayat raj* that James Bruges describes. The main difference between my proposal and his is that I believe funding decisions, including those which require pooling of money, should primarily be made by individuals acting within small community groups rather than by elected officials acting on their behalf.
25. These emergency funds could eventually be replaced by a permanent basic income scheme as described further back.
26. There is nothing fantastical about the idea of 'underdeveloped' people donating to help others, including supposedly more developed communities, in times of crisis. At the time of the Irish famine in the nineteenth century the impoverished Native American Choctaw tribe, who had themselves experienced a famine a couple of decades earlier, made a donation to Ireland. Probably the most recent example is the donation of $50000 from the city of Kandahar in Afghanistan to the Japanese relief effort in the aftermath of the 2011 earthquake in Japan.
27. Esquivel *et al* (2010), "A Decade of Falling Inequality in Mexico: Market Forces or State Action?", UNDP
28. ibid, p 26
29. Soederberg, Susanne (2010) 'The Mexican competition state and the paradoxes of managed neoliberal development', *Policy Studies*, 31: 1, 77 – 94
30. Rosenberg, Tina, "To beat back poverty, pay the poor", *New York Times*, January 3 2011 http://opinionator.blogs.nytimes.com/2011/01/03/to-beat-back-poverty-pay-the-poor/
31. "Family-friendly: Brazil's scheme to tackle poverty", BBC, 25 May 2010 http://www.bbc.co.uk/news/10122754
32. "Brazilians upbeat about their country despite its problems", Pew Research Center, September 22 2010 http://pewglobal.org/2010/09/22/brazilians-upbeat-about-their-country-despite-its-problems/
33. A notable exception to this rule is Mexico. Violent crime there has risen sharply in recent years, despite the success of the Oportunidades social transfer programme. This increased crime rate appears to derive from extensive drug trafficking between Mexico and its much wealthier neighbour, the USA, with whom it shares a long border. The violence is concentrated in areas of the country that are strongly affected by the drug trade.
34. Wilkinson, Richard and Pickett, Kate (2010) *The Spirit Level: Why equality is better for everyone*, Penguin
35. "For the Executive With Everything, a $230,000 Dog to Protect It" New York Times, June 11 2011 http://www.nytimes.com/2011/06/12/us/12dogs.html?pagewanted=1&_r=1
36. Shaxson, Nicholas (2011) *Treasure Islands: Tax Havens and the Men who Stole the World*, Bodley Head

Chapter 8

Reasons for optimism

Richard Douthwaite
with David Knight

First we discuss the interlinked problems of climate change, peak fossil fuels and the credit crunch and then grounds for some optimism , including means of adjusting energy and commodity markets to start to address these ills, and other measures to deal with non-CO_2 greenhouse gas emissions.

The world has warmed by approximately 0.7 degrees Celsius in the 200 years since fossil fuels began to be used on any significant scale[1]. The rate at which the planet is heating up could be accelerating and 2010 was the hottest year since the global record began, 131 years ago[2]. All ten of the warmest years on record have occurred since 1998.

The warming has not been uniform. The biggest temperature rises have been around the North Pole where some worrying self-reinforcing feedbacks have developed. For example, the Arctic ice has been melting unexpectedly rapidly. In December 2010 Arctic sea-ice cover was the lowest ever recorded, with an average extent over the month of 12 million square kilometres, 1.35 million square kilometres below the 1979-2000 average for December. In 2011 the September Arctic sea-ice cover was the second lowest recorded. The melt could be increasing the rate at which the planet heats up because the white ice which reflected solar energy back into space has been replaced by dark, heat-absorbing sea. Similarly, the warming is melting the permafrost in Russia and releasing large amounts of methane, a powerful greenhouse gas, into the atmosphere.

Illustration 1

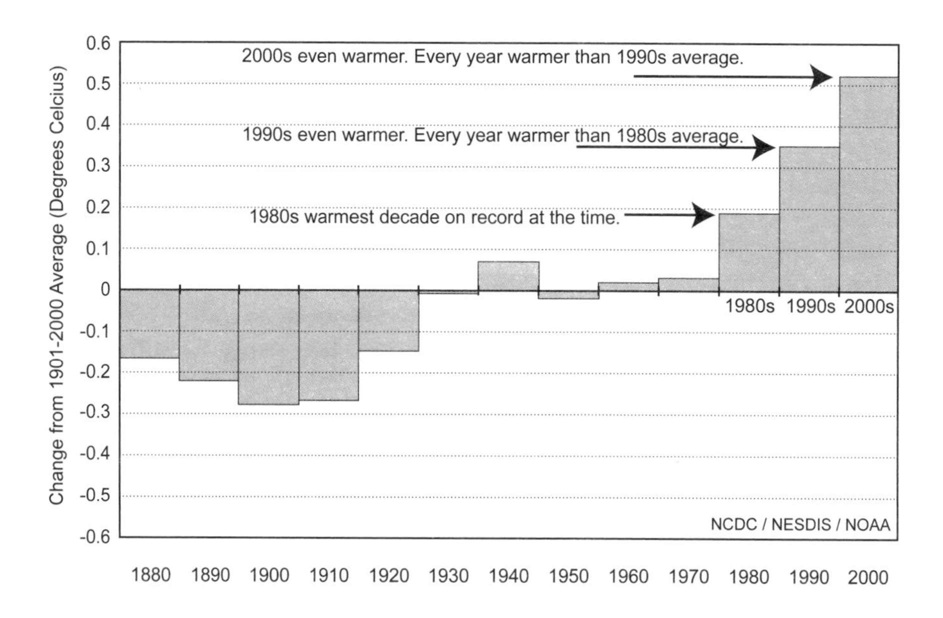

How the world's average temperature has changed from decade to decade. The rate of increase in the last three decades has been extremely rapid.

Even if the atmospheric concentration of greenhouse gases was to be stabilised at its current level, the Earth would continue to warm by at least another 0.7 degrees. This is because the full heating effect of these gases takes some years to overcome the Earth's considerable thermal inertia. This extra warming would obviously melt more ice, release more methane and accentuate other feed-backs as well.

The public debate about what should be done to stem the warming has focussed on just one greenhouse gas from just one source – carbon dioxide emissions from fossil fuel use. However, there is only the remotest possibility that the international community will take effective action through the existing United Nations' mechanisms to reduce fossil fuel CO_2 emissions in the limited time left before the feedbacks become self-reinforcing and generate a rapid, unstoppable and disastrous shift to a very much warmer climate. This is because reducing the amount of CO_2 released from the burning of fossil fuel can be achieved in only three ways. One is to burn less fossil fuel but this reduces the amount of energy the economy has available which in turn tends to reduce people's incomes. As no society wants its incomes to fall, a lot of effort has been put into burning fuel with increased efficiency and into developing non-fossil energy sources to try to break the historically-close link between income and fossil-fuel use.

So far, however, only a limited number of countries have managed to continue to grow their economies while reducing their fossil fuel use. Sweden is one. It reduced fossil fuel's contribution to its total energy supply from around 80% in the early 1970s to about a third today[3] by increasing its production of nuclear (Just under half its electricity comes from nuclear plants) and renewable energy while at the same time stabilising the amount of energy used per head of population. It also increased the efficiency with which it used energy, reducing the amount required to produce $1,000-worth of output by a third.

Unfortunately, however, the Swedish model cannot be applied to more than a handful of countries. Nuclear energy is not fossil-fuel free and a long time is required for a reactor to recoup the energy input and CO_2 emissions incurred in construction. This means that reactors cannot be built fast enough on a world scale to cut CO_2 emissions at the rate dictated by the latest climate science. Moreover, even if it made sense to temporarily boost CO_2 emissions and drastically reduce the supply of energy and capital to the rest of the global economy to allow the plants to be built, there would be insufficient high grade uranium ores to run them[4] and mining what little there is, fossil-energy intensive.

Thorium is being touted as a more-abundant alternative to uranium but its use has not yet been demonstrated in a commercial reactor and by the time one is built, the time left for decisive climate actions to halt the feedbacks may well have run out. Nor is producing plutonium in fast breeder reactors a solution to the uranium supply problem. The required reprocessing of fuel to separate plutonium, uranium and fission products is not cost effective without hefty subsidies from nuclear weapons programs. Moreover, the large scale plutonium economy needed for a world-wide fast reactor programme brings with it severe risks from nuclear proliferation and all forms of nuclear terrorism. Renewable energy sources will not enable us to make massive emissions cuts either as they also take time and require a lot of fossil energy to develop on the required scale.

Because we have left making the transition to renewable energy so late, using a lot less fossil fuel would mean that less energy would be available in total and that, as a result, incomes would be lower than today's for the foreseeable future. Governments are reluctant to accept this, of course, and have been hoping that an alternative way being proposed to reduce fossil carbon emissions will enable their countries to continue to burn fossil fuels with a clear conscience on a scale not too much less than their current one. This involves capturing the carbon dioxide produced when the fuel is burned and locking it safely away for many thousands of years, perhaps by pumping it into disused gas wells. Unfortunately, the process requires a lot of energy itself and thus increases the demand for fossil fuels.

Illustration 2:

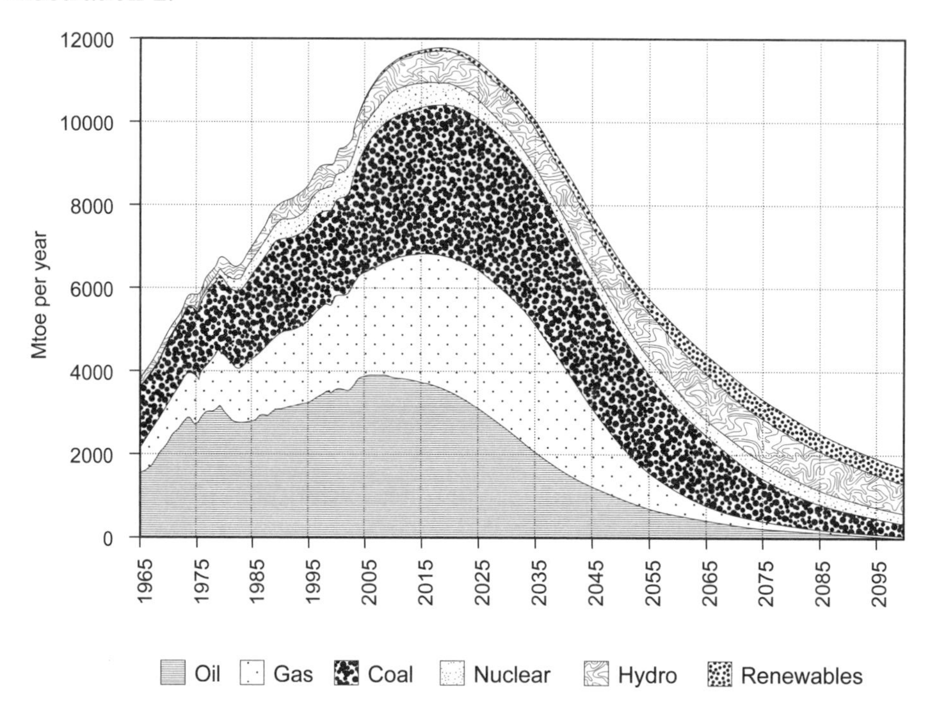

Paul Chefurka's estimates indicate that the total amount of energy available to the world will begin to fall about 2020 and could be half the maximum level thirty years later. This is because supplies of nuclear, hydro and renewable energy cannot be developed fast enough to make up for the rate at which fossil output declines.

These themselves will be in short supply. The world's supply of oil has been flat since 2004 because the producers have been unable to bring new sources on line faster than the output from their older fields has declined. Their inability arose because oil is getting harder to find and extract – BP would not have attempted to open its disastrous well 1,600 metres below the surface of the Gulf of Mexico if it had had any better options. The increasing production difficulties mean that the supply of oil will soon begin to decline and that, year by year, the decline will be at an accelerating pace.

Although the output of coal and gas is still increasing, their supply will begin to contract too in a few years' time. A Canadian analyst, Paul Chefurka, expects[5] coal output to reach its peak in about five years and conventional natural gas in about 15. He omits shale gas supplies from his projection as the extent of its future is very unclear, as we will see. For the reasons we just noted, he does not expect the supply of renewable and nuclear energy to expand fast enough to compensate for the decline in these fuels and, as a result, he believes that humanity's total energy supply from all sources will begin to decline after 2020. The decline he expects is shown in Illustration 2, which also indicates that he believes there will be very little growth in the overall energy supply before the decline begins. This implies very little increase in world incomes over the next decade.

A similar estimate[6] of the global energy supply was prepared for the 2007 Zero Carbon Britain report. This showed a gentle decline in output after 2010 and a more rapid fall after 2025. Similarly, the most recent forecast[7] by the oil and gas geologist Colin Campbell, one of the founders of the Association for the Study of Peak Oil, indicates that the total amount of energy available from oil and gas production will decline slightly between now and 2020 and then begin a more rapid decline. The industry is already admitting difficulties. Oil and gas supplies will struggle to keep up with world demand growth, Peter Vosser, the chief executive of Shell, told *The Financial Times* in Sept 2011. "The world needs to add the equivalent of four Saudi Arabias or ten North Seas just to *keep the supply level*," he added. [My italics]

Coal will be in short supply too. In a paper[8] published in May 2010, two American academics, Tadeusz W. Patzek and Gregory D. Croft conclude: "The global peak of coal production from existing coalfields is predicted to occur close to the year 2011... [T]he peak carbon emissions from coal burning are 4.0 Gt C (15 Gt CO_2) per year. After 2011, the production rates of coal and CO_2 decline, reaching 1990 levels by the year 2037, and reaching 50% of the peak value in the year 2047. It is unlikely that future mines will reverse the trend predicted in this business-as-usual scenario." Other observers put the peak later. For example, Mikael Höök from Uppsala University and his colleagues[9] expect it between 2020 and 2050. However, they conclude: "Global coal production could reach a maximum level much sooner than most observers expect."

One reason for this earlier peak could be that the Energy Return on the Energy Invested (EROEI) in developing new coal mines and the infrastructure required to get the coal to where it will be burned is very low. A study[10] carried out by Jamie Bull, a sustainability consultant based in London, puts it at 5.5. In other words, if you invest one unit of energy in developing a coal mine and its infrastructure, you get 5.5 times that energy back over the mine's life. That might seem a good return but in fact it is not. New oil fields are giving perhaps three times that return – and, when easy oil was available, as it was in Texas in the 1930s, they had EROEIs of around 100 – that is, 18-20 times better.

You might think that it would be worth producing coal for energy use so long as the return was better than 1 for 1 as this would give you back slightly more energy from the coal than it took you to extract it. However, if that return applied to all a country's energy sources it would mean that the economy devoted almost all its efforts to getting its energy and had almost no energy left for doing anything else. Professor Charles Hall of New York State University, who developed the EROEI idea after investigating why fish used energy to migrate, believes that, in view of the amount of energy it takes to run our civilisation, its energy sources need an average return significantly higher than 3:1 for it to persist.

If this is correct, it means that new coal production is already a marginal energy source for our society and its output might decline very quickly particularly as the return on other forms of energy investment are much better. Hall calculates that wind power has an EROEI of 25 and, in places where there is a big difference between high and low tide, barrages can give back 115 times more energy over their life than they take to build. By contrast, if the carbon dioxide from burning coal has to be sequestered by compressing it and pumping it into a disused gas well, he calculates that only 1.5 times the invested energy comes back. This is far too little for carbon capture and storage (CCS) to be a commercial possibility.

Indeed, the tide against coal has already turned in the US according to Lester Brown. He writes[11]:

> Utilities are beginning to recognize that coal is not a viable long-term option. TVA [mining corporation] announced in August 2010 that it was planning to close 9 of its 59 coal-generating units. Duke Energy, another major south eastern utility, followed with an announcement that it was considering the closure of seven coal-fired units in North and South Carolina alone. Progress Energy, also in the Carolinas, is planning to close 11 units at four sites. In Pennsylvania, Exelon Power is preparing to close four coal units at two sites. Xcel Energy, the dominant utility in Colorado, announced it was closing seven coal units. And in April 2011, TVA agreed to close another 9 units as part of a legal settlement with [the] EPA [Environmental Protection Agency]. In an analysis of the future of coal, Wood Mackenzie, a leading energy consulting and research firm, describes these closings as a harbinger of things to come for the coal industry.

The situation with gas is harder to read. The rate at which global gas fields were being discovered peaked around 1970 and has declined so sharply since with the result that consumption has outpaced discovery since 1980. This would lead one to predict that production would peak soon and that the rate at which output declined would be even faster than for oil as this is characteristic of individual gas well depletion. But the development of shale gas supplies invalidates this prediction. Shale gas is produced by blasting a cocktail of chemicals and other materials into gas-containing rock to shatter it into myriads of pieces. The cracks this creates allow the gas to be pumped out. Environmental groups and the US Environmental Protection Agency are worried about the effects that this fracturing and the chemicals used will have on drinking water supplies and on the environment generally. Very large volumes of water are pumped down the well during fracking and, even if no noxious chemicals are added to it, when it comes back to the surface it could be heavily contaminated with arsenic and other heavy metals which have leached out of the shattered rocks though which it passed. A slurry of rock fragments is discharged, too. Both the slurry and the liquid need to be disposed of safely. This is expensive so the conventional approach is just to run it into a lagoon with the promise of remediation later on. As a result of these concerns, New York State has imposed a moratorium on drilling while New Jersey has banned it permanently. So has France. Bans have also been imposed in parts of South Africa and Australia. The industry, however, denies that groundwater supplies are being polluted because the fracturing occurs far below the level from which the water is extracted.

From a climate point of view, the process also looks unpromising because some of the gas, which is mainly methane, a very powerful greenhouse gas, escapes into the atmosphere. As a result, a Cornell professor, Robert W. Howarth, estimates[12] that that shale gas may be slightly worse for the climate than coal. Tom Wigley of the US National Center for Atmospheric Research, Boulder has recently confirmed[13] Howarth's finding. If the warming effect of methane at 20 years is taken to be 72 times that of CO_2 [14] and Howarth's upper bound figure of 7.9% [15] for the percentage leakage of methane during north American shale gas production is right, it is easy to calculate the warming effect of the methane is about twice as great as that of the CO_2 produced by burning the shale gas without carbon capture and storage.

Shales suited to fracking are found all over the world and a lot of gas could be produced from them provided that the environmental objections can be overcome and the EROEI is sufficiently high for the process to compete with renewable energy technologies. So far, however, the EROEI figure does not look good – it could be 4 or 5 – because the amount of gas released by the fracturing starts to decline quickly after a well is opened, giving each well a relatively short useful life. A well can be re-fracked up to five times after its first decline, perhaps every four to five years for successful wells[16] but it is not clear how fast production will decline after each re-fracking. As a result, it will be some time before the EROEI of the whole cycle is known from sinking the well to reaching the consumers. Early financial returns may look so promising that investors pile in but the long-term energy return may disappoint.

Nevertheless, taking oil, coal and gas together, we know enough to conclude that emissions from fossil energy are going to fall anyway without any international agreement and people's incomes will fall with them. The trick that governments need to pull off is to manage the income decline so that a catastrophic economic collapse is avoided. To do this they need to understand why the world economy almost collapsed after the Lehman Brothers bank failed in 2008.

Because the supply of oil had not increased since 2004, its price went up and up in the period before the bank failed, as demand for it grew, taking the prices of gas, coal, food and other commodities with it. These increasing prices for energy and commodities meant that more and more money had to leave the net importing countries to pay for these imports. A lot of the money spent this way was not returned in the form in which it left, to the countries that spent it. It went out as income and came back as capital. I'll explain. If I buy petrol for my car and part of the price goes to Saudi Arabia, I can only buy petrol again year after year if that money is returned year after year to the economy from which my income comes. This can happen in two ways, one of which is sustainable, the other not. The sustainable way is that the Saudis buy goods and services from the country in which I live or from countries from which my country imports less than it exports. If the Saudis do this, the money returns to my country as income. The unsustainable way is that the Saudis lend it back to my country. This returns it as capital. The loan enables my country to continue buying oil, but only by getting deeper and deeper into debt which will eventually need to be repaid.

As commodity prices rose in the years before 2008, the flow of money to the energy and mineral producers increased so much that there was no way that the countries concerned could spend it all back into circulation. They lent it back instead and eventually the level of debt in the consumer countries became too great to be borne. In the US, as people sought to maintain and improve their lifestyles by borrowing, the ratio of household debt to household income grew by more than 40% to around 130% between 1997 and 2007. In Britain, the ratio rose by over 50% to a similar 130% and Ireland outdid both US and UK, increasing its ratio by 85% to a massive 190%.

The weakest US borrowers, those with sub-prime mortgages, were the first to be unable to pay and, because of the way their loans had been "securitised" and sold off around the world, no-one knew which banks were nursing securities which had lost most of their value. The consumer countries' banks lost faith in each other and, had governments not stepped in with bailouts and guarantees, many would have collapsed.

So the fundamental reason for the "credit crunch" and the over-indebtedness of the consumer countries was that the commodity suppliers could not expand their output because, for example, oil was getting harder and harder to find and production in many fields declining, and this enabled them to put their prices up to such an extent that their customers' economies eventually collapsed. Until the role of energy supply constraints is recognised and corrected, the global economy will never function well again because the moment that the "green shoots of recovery" appear in the consumer countries and spending begins to rise, their demand for energy will go up. It will quickly exceed the available supply and the price they have to pay for it will soar. The money they use to pay the higher prices will not be available for domestic spending. As a result, the green shoots will wither unless the energy exporters return all their increased income by purchasing goods and services from their customer-countries. If the energy exporters save at all, (and they will want to, because they know that they are exploiting a depleting resource) they will risk undermining the attempted recovery; the value of their investments in their customers' economies falls as the recovery falters, as does their own income stream as prices fall back.

Correcting the way the energy market works is therefore in the interests of both buyers and sellers. The required correction involves sharing out what economists call the "scarcity rent" – the extra over and above production costs that producers are able to charge when something gets scarce. At present, all the scarcity rent is kept by the producers with the result that debt and income-distribution problems build up. The sharing must therefore involve limiting the amount of rent the producers receive, and distributing the remainder in a way that ensures their customers can keep consuming.

Indeed at least one meeting of the International Energy Agency (IEA) has discussed the possibility of establishing a cartel of the net energy importing nations to act in their joint interest to limit the price paid to producers, and thus their share of the scarcity rent[17]. Such a cartel could conceivably operate as follows:

- It would agree with the fossil energy producers how much coal, conventional and unconventional oil, and gas from any source they would supply at a fixed price each year for the next five years. The agreed price would probably rise in successive five year periods in real terms as output fell. In return, the producers would undertake only to supply fuel to buyers with purchase permits from the cartel.
- Each month, the cartel would auction permits, with separate permits entitling the purchaser to buy specific quantities of one of the three fuels. Provided the global economy was buoyant enough for fuel demand to exceed the production level agreed with the producers, the price that bidders offered for permits would be above zero and the cartel would earn an income, very probably a large one, from their sale.
- The permit income would then be shared amongst all the members of the cartel on a previously-agreed basis.

These arrangements would have much in common with the Feasta Climate Group's proposals for Cap and Share. Such a scheme would keep as income most of the income used to buy energy – only the savings that the energy producers were able to make out of their guaranteed incomes would become capital, a lot of which would be likely to be spent on construction projects in consumer countries, converting capital straight back to income again.

The advantages of the arrangement would be enormous. The cartel's guarantee of high and rising real prices would give the producers a better return from keeping their oil, gas and coal in the ground than they could be sure of getting if they extracted them and invested the proceeds on Wall Street. The avoidance of oil and other energy price spikes would help to help to stabilise the global economy. Although the total price of energy including the cost of the permit would still be set by the market, renewable energy producers would be able to invest with confidence as they would know what the base price of fossil energy was going to be for five years ahead. The fossil fuel energy market itself would be buoyant and stable because the investment in renewable energy would boost the economy, paradoxically encouraging an increased demand for fossil fuels at least in the early cycles before the production of renewable energy took off.

How would the cartel's version of Cap and Share differ from the Feasta Climate Group's version? In the Feasta version, the international community would set up a special agency – let's call it the Global Climate Trust – to handle CO_2 emissions. Taking the best scientific advice, the Trust would decide on the rate at which emissions from fuel use needed to fall year by year to prevent the climate feedbacks from becoming uncontrollable. Many climate stabilisation proposals produced in line with the latest science envisage cutting fossil fuel CO_2 emissions by between 80% and 100% by 2050, and if this was the reduction the Trust adopted, an annual roll-back rate of around 6% would be required. At that rate, emissions would be reduced by a third after seven years, by a half after twelve, by two-thirds after 19 and would be down to 10% of their present level by 2050.

The cartel, on the other hand, would not impose for climate reasons an annually-declining cap on the amount of fossil energy it agreed to buy from the producers. Its limit would be set by what the producers contracted to supply at the negotiated price. But this difference might not matter very much because the rate at which oil production is likely to decline as a result of production constraints could be at around the 6% rate anyway. Indeed, the IEA reported in 2008 that production from 600 existing fields was already declining at 6.7% a year, far more than the 3.7% decline it had estimated in 2007. However the IEA World Energy Outlook 2010 considered that the decline in crude oil production from existing wells will be more than compensated for by increased production of crude from reserves yet to be exploited and reserves yet to be discovered together with increases in natural gas liquids and other types of unconventional oil and that total oil production will continue to rise to 2035. In contrast, most recent reports on future oil supplies such as that published by the British Government-funded Energy Research Centre in October 2009 and by Kjell Aleklett's group have presented strong evidence that the IEA's analysis is badly flawed, concluding instead that total world oil output will peak by 2020[18].

But while emissions from producing and burning conventional oil, coal and conventional natural gas output might decline under the cartel arrangement at about the rate the Feasta Global Climate Trust might require, there are three reasons why this may not be fast enough to prevent dangerous global warming. We have dealt with one – shale gas production. The other two are the possible massive development of the Canadian tar sands possibly using steam produced directly by nuclear reactors to melt the tar; and the underground gasification of coal. The latter involves drilling two boreholes into a coal seam and linking them, either by drilling horizontally or by fracking, exactly as with the shale gas. Air and/or steam is passed down one hole, the coal is set alight and the gases from its controlled, partial combustion emerge from the second hole. They consist of hydrogen (especially if steam is injected as in the conventional production of water gas), carbon dioxide, carbon monoxide, methane, and small quantities of ammonia, oxides of nitrogen and, depending on the sulphur content of the coal, hydrogen sulphide. The latter has to be removed but the rest of the gas can be burned directly in a gas-fired power station, which, it is claimed, could be fitted with carbon capture equipment so that the CO_2 in the exhaust fumes could be pumped back into the coal seam from which the carbon came after the burning operation has moved on.

If this sequestration was actually done – and doing so would obviously involve extra costs – there would still be two serious environmental drawbacks. One is subsidence– when the coal is burned out, the rocks above collapse and the land sinks, causing structural damage to buildings on the surface. This is a well-known problem in conventional coal-mining areas and mines would frequently avoid digging coal from under important buildings to avoid having to pay substantial sums in compensation to their owners. The other threat is to aquifers which can become polluted by the carcinogenic tars left by the burning coal, particularly as the rock strata around the burn break up.

Underground coal gasification has been used in the past but most projects seem to have been abandoned because low gas and oil prices made them uneconomic. However, higher energy prices in future are likely to lead to a massive resurgence, particularly if environmental controls are lax.

Thus the large size of coal, shale oil , shale gas and tar sand reserves and the ease with which these can be exploited by the methods outlined above have potentially grave consequences for atmospheric CO_2 levels and temperature rises [19]. It follows that there is no guarantee that reductions in the supply of fossil fuels from conventional sources as a result of resource constraints will directly prevent catastrophic temperature rises. It is therefore important that, if an energy-consumers' cartel is established, it has climate change goals. This is potentially not a problem because the cartel would be set up by governments rather than short-run profit-maximising multi-national corporations. However, it is unlikely that climate goals will be built into the cartel's constitution unless there is strong pressure from the electorate to do so. The low probability at present that this pressure will be forthcoming is a flaw in this approach.

The other area in which the consumer cartel might produce an inferior result to the Feasta Climate Group's version of Cap and Share is in the way it would distribute the money it raised when it auctioned the permits giving purchasers the right to buy fossil fuels from the producers. The Feasta's Global Climate Trust would auction permits too and would need to distribute its income in five ways:

1. Compensation payments for higher energy prices.
2. A Carbon Maintenance Fee to protect soil and forest carbon stocks
3. A Hardship Fund for communities particularly hard hit by the effects of climate change or the transition to non-carbon energy.
4. Investment capital for renewable energy development.
5. The operating costs of the Global Climate Trust itself.

I will comment on the first four in turn and then we will see whether the cartel might need to do much the same.

1. **Compensation payments.** Fossil fuel producers will have to pass on the cost of the permits they buy to their customers. As the supply of permits gets increasingly tight, their price, and thus the price of energy will go up, taking the price of food and other necessities with it. People will have to be compensated individually for these rises at some basic level as, otherwise, the poor would be driven out of the market and go cold and hungry. However, the basic compensation can never be enough to cover the increase in the cost of living of people whose lifestyle involves the direct or indirect use of a lot of energy. This group, which is made up of richer people in every country, will inevitably find that their cost of living goes up by more than their compensatory payment.
2. **The Carbon Maintenance Fee.** The higher energy prices will put extra pressure on the world's forests and pasture land. Unless these were protected they would be opened up for biofuel production and massive amounts of carbon dioxide would be released and other ecosystems services badly damaged as a result of the land-use change. This is already happening in Indonesia where it is reckoned that when jungle growing on peatland is cleared for palm oil production, the amount of CO_2 released is so great that it could take over a century for it to be recovered by burning the palm oil instead of diesel fuel[20].
   The payment of a Carbon Maintenance Fee (CMF) to reward countries which maintain the amount of carbon locked up in their soils and bogs and in the plants growing on them is therefore an essential part of any method of limiting fossil fuel CO_2 emissions. Under the CMF arrangements, a small sum would be paid annually for each tonne the country concerned had kept intact, the latter determined as a proxy

based on the numbers of hectares of each particular land use type. The payment would cover the opportunity cost the country was carrying by refraining from putting the land involved to a more profitable use. However, if the tonnage being safeguarded fell, the country responsible would have to buy special emissions permits[21] to cover the amount of carbon involved. On the other hand, if a country increased the amount of carbon sequestered in its biomass and soils, it would get the special permit price for every tonne. I will discuss the Carbon Maintenance Fee scheme in more detail later on.

3. **The Hardship Fund.** People in some parts of the world will be especially hard hit by the effects of climate change – those in Bangladesh or the Maldives threatened by rising sea levels, for example. Other communities may need to make exceptional adjustments to cope with much lower levels of fossil fuel use. All countries, rich and poor, would be able to claim both types of assistance from this fund. Assessing these claims and allocating funds fairly would be possible but not easy.
4. **Energy investment.** There will be calls for a lot of the Trust's annual income to be used to fund the development of non-carbon energy sources and an international fund is likely to be set up to make loans to multinational projects such as those involved in the development of concentrated solar power in the Sahara, and to national governments for their own investment in energy projects. However, it is important that communities and families can make energy investments too and part of the Trust's income should be allocated for this purpose.

The Trust would almost certainly pay part of the money it collected each year from the sale of permits on an equal per capita basis to everyone on the planet to compensate for higher energy and food costs (see 1. Above). Most of the remainder would go to governments under the Carbon Maintenance Fee, Hardship Fund and Energy Investment headings. The residue would be retained to finance the Trust's operations which would include programmes to deal with the non-CO_2 contributors to the warming effect I will mention later on.

There would be pressures for the cartel to spend its auction receipts in a similar way. For example, the governments of the poorer countries in the cartel would insist that part of the money came to them on a population-related basis because they needed to deal with the increase in fuel-poverty that the scheme would inevitably bring about. Some money would probably be used to pay a Carbon Maintenance Fee, some would go to a hardship fund and the remainder might be placed in a revolving loan fund to finance projects to reduce fossil energy requirements against the day they ran out.

However, the big, important difference between the Trust's distribution and that by the cartel is that much more of the Trust's money would go directly to people whereas the cartel's would go to its member governments and it would be left up to them to decide how much of it they passed on to their citizens to spend. This difference arises because the Feasta Climate Group regards the right to a share of the benefits of using fossil fuel as a human right, one that belongs to each individual, rather than a national right. As a result, a very high proportion of the Trust's income from permit sales would go to families and, since collective action on energy can be more effective than family-level action, to the communities in which they live. Governments, however, are unlikely to see things this way. They believe that they can make better decisions than those that emerge from the "wisdom of crowds" and they would want to maximise their control over the cartel money. It is therefore important that a Trust is established to promote the true Cap and Share before, say, the G-20 governments begin to discuss an ersatz version which pays out solely to governments and which has no inbuilt climate objectives, being largely concerned with getting the global economy working again..

Action beyond CO_2

If fossil carbon dioxide emissions are going to fall rapidly as a consequence of the introduction of a cap (and they would fall even more quickly but with terrible human consequences if the world economy collapsed because something akin to Cap and Share was not introduced) those concerned about the climate can move on to deal with other aspects of the crisis. This will be both productive and liberating.

It is not generally recognised that the fossil CO_2 we've been so bothered about may have contributed as little as between 7 and 20% to recent warming if the feedback effects of water vapour, itself a greenhouse gas, are included. However, as there is a lot of uncertainty about the role played by water vapour, it is usually ignored, particularly as its concentration at a given temperature is limited as it can only increase until the atmosphere becomes saturated whereupon it forms clouds and falls as rain or snow. The uncertainty about the effects of water arises because, although higher temperatures lead to more water in the atmosphere, the extent of cloud formation is difficult to predict. If the water stays as a vapour, it blocks heat radiation into space because it is opaque to certain infra-red wavelengths. On the other hand, if it forms clouds, these have a net cooling effect because their whiteness reflects more heat back into space during the day than they prevent escaping by acting as a warming blanket at night.

Cloud formation may be a rather complicated process. Recent work[22] has indicated that the bacterium *Pseudomonas* floating in the atmosphere can seed ice crystal formation[23] and in so doing may play an important in forming the low-level cloud that is particularly effective in keeping the Earth cool by reflecting solar radiation. As the bacterium grows on the leaves of trees and is blown into the atmosphere from them, the planting of forests could have a much greater effect on the Earth's temperature than the amount of carbon they extract from the atmosphere would indicate. So this is a feedback we might be able to usefully exploit. In addition, minute single-celled plants (phytoplankton) growing in the top layer of the sea may also have an important effect on cloud formation, in this case by producing a chemical called DSP [24]. The DSP breaks down to DMS which is released to the atmosphere where it breaks down further yielding sulphur dioxide. This joins that produced by volcanoes and the burning of sulphur-containing coal and oil. The sulphur dioxide gives rise to sulphate aerosols encouraging the formation of the water droplets which form clouds. Thus maintaining the production of marine phytoplankton or even enhancing it may help to combat global warming.

Table 1

Gas/pollutant	Heating effect of rise in concentration 1750-2005[a] (Watts per square metre)	% share of the total heating effect	Heating effect of rise in concentration 1990-2005 (Watts per square metre)	% share of heating effect 1990-2005
Carbon dioxide	1.66 ± 0.17	48.4	0.364[b]	68.04
Methane	0.48 ± 0.05	14.0	0.023[b]	4.30
Nitrous oxide	0.16 ± 0.02	4.7	0.033[b]	6.17
Halocarbons	0.34 ± 0.03	9.9	0.044[b]	8.22
Low-level ozone	0.35 (-0.1, +0.3)	10.2	0[c]	0.00
Black carbon[d]	0.44 ± 0.25	12.8	0.071[e]	13.27

Sources:

a from Section 2.9.3 Global Mean Radiative Forcing by Emission Precursor, IPCC Fourth Assessment Report (AR4)

b NOAA downloaded from http://www.esrl.noaa.gov/gmd/aggi/

c Tropospheric ozone's uneven geographical distribution and marked variation on a fairly short timescale make it difficult to determine trends in its distribution. However there appears to have been little net change in total column ozone concentrations over this period at least in the Arctic, although values have fluctuated markedly [25].

d The figures for black carbon are based on the IPCC value for the direct heating effect which may be an underestimate (see below) [26].

e This figure is only for black carbon derived from the combustion of fossil fuel and biofuels and does not include open biomass burning, e.g. from clearing forests, which is very difficult to estimate [27].

For the moment, however, if we leave water vapour, possible changes in solar activity and an increase in the cooling effect due to sulphate aerosols out of consideration, CO_2 was responsible for some 48% of the warming that has happened since 1750 as Table 1 shows. But estimates[28] from the Woods Hole Research Center in Table 2 show that during the 1990s, about a quarter of the CO_2 released came from deforestation and land use change such as the ploughing of grasslands.

Table 2:

Carbon emissions sources and sinks Average annual figures for the 1990s	Billion tonnes carbon	Margin of error in bn. tonnes	Billion tonnes CO_2	Margin of error in bn. tonnes
Emissions from burning fossil fuel	6.3	±0.4	23.3	±1.5
Emissions from changes in land use Deforestation (85%) Cultivation of prairie soils (15%)	2.2	±0.8	8.1	±3.0
Less absorption by oceans	2.4	±0.7	8.9	±2.6
Less absorption by unknown sinks	2.9	±1.1	10.7	±4.1
Annual increase in the atmosphere	3.2	±0.2	11.8	±0.7

Only rough figures are available for the tonnage of carbon released into the atmosphere and even less reliable estimates can be given for where that carbon goes. Some researchers refer to a "missing sink" because of their ignorance about where perhaps a third of the emitted carbon actually ends up. They suspect that it is being taken up by plants and soils but this has not been verified. Source: Woods Hole Research Center, http://www.whrc.org/carbon/index.htm and CCSN.

This means that CO_2 from fuel use contributed about 50% to the warming effect during that decade. This was more than any other single source. However, since the runners-up also contribute substantially – methane with approximately 4%, nitrous oxide 6%, CO_2 from deforestation and land use change about 10% and halocarbons about 10% – it is clear that, although action is necessary to reduce the rate at which CO_2 from fossil fuel use is being added to the atmosphere, these smaller emissions sources must be dealt with as well and need their own reduction programmes. Black carbon formed by the incomplete combustion of fossil fuels, biofuel, and biomass is of particular importance in this respect. Although it contributed about 13% of the warming effect in both 1750-2005 and 1990-2005 periods (see Table 1), it had the third largest heating effect in the former period but was second largest in the latter. This change in ranking resulted from the negligible overall change from 1990-2005 in atmospheric methane concentrations both in total atmospheric column[29], and surface measurements[30]. This analysis should not encourage complacency about methane as surface atmospheric concentrations have started to increase again from about 2007[31]. But the importance of black carbon may be even greater than implied by the figures in Table 1. This is because the figures here are based on the IPCC AR4 value of 0.34 W/m^2 for the direct heating effect (globally averaged direct radiative forcing) of black carbon. This may be a considerable underestimate; the IPCC's value is much smaller than the 0.58 W/m^2 obtained by averaging the values from the seven most recent studies[32] which in the highest case calculated that the value could be as much as 1.0 W/m^2.

Removing CO_2 from the atmosphere

The worryingly rapid rise in temperatures since 1980 shown in Graph 1 indicates that the collective warming effect of all the atmospheric constituents is already too great. Although two Carnegie Institute researchers, Long Cao and Ken Caldeira, have calculated[33] that for every 100 billion tons of carbon removed from the atmosphere, average global temperatures would drop 0.16°C, no-one knows with any certainty how much CO_2 needs to be removed and sequestered safely away before warming stops. Indeed, that question cannot be answered at all. This is not just because of the level of scientific uncertainty but also because reductions in the other gases' concentrations would mean that less CO_2 would need to be removed from the atmosphere.

However, it has been suggested that about a tenth of the current atmospheric tonnage should be removed on a "let's see if that's enough" basis. This would involve bringing the atmospheric concentration down from its present level of about 389 parts of CO_2 per million (ppm) to 350. This 350 target is being promoted by a well-run NGO, 350.org, and has gathered support around the world. For example, Lord Stern, a former chief economist at the World Bank and author of the *Stern Review*, a 2006 report to the British Government on the cost of tackling climate change, told a German journalist[34] in September 2009 that he thought that 350 was "a very sensible long-term target."

The leading American climate scientist, James Hansen[35] sees 350 as an interim goal to be re-assessed in the light of new data. He thinks it may not be enough because the melting of the Arctic sea-ice began in the 1970s before the 350 level was reached and suggests that 300-325 ppm might be necessary to maintain the ice. Professor John Schellnhuber, the director of the Potsdam Institute for Climate Impact Research in Germany, thinks that almost any concentration above the pre-industrial level of 270 ppm might be too much. He told The Guardian in 2008[36] that even a small increase in temperature could trigger several climatic tipping points.

"Nobody can say for sure that 330 ppm is safe," he was quoted as saying. "Perhaps it will not matter whether we have 270 ppm or 320 ppm, but operating well outside the [historic] realm of carbon dioxide concentrations is risky as long as we have not fully understood the relevant feedback mechanisms."

To return to the 350 level would require the sequestration from the atmosphere of a quantity of CO_2 equal to all that added as a result of human activity since the 350 level was passed in 1987 together with all the CO_2 that will be released before all fossil fuel use stops. Some, perhaps all of the CO_2 that has been absorbed by the sea since 1987 will need to be recovered too because it will be released again as the atmospheric concentration falls. However, this "out-gassing" will take place relatively slowly and Long Cao and Ken Caldeira estimate that its recovery could be spread over 80 years[37].

Certainly, the sea should no longer be regarded as an acceptable sink for CO_2 emissions as the acidity of its water is rising in step with the amount of CO_2 it contains and this is threatening animals and phytoplankton with calcium carbonate shells or skeletons. These make up more than a third of all marine life. Core samples taken from coral reefs show a steady drop in calcification over the last 20 years and it is feared that if CO_2 levels in the atmosphere were to reach about 500 parts per million, the resulting rise in the seawater concentration would cause calcification to cease.

The recognition that all future fossil fuel carbon dioxide released to the atmosphere from now on and almost all CO_2 emissions for perhaps the past three or four decades will have to be recovered would have a significant bearing on the Trust's decision on how rapidly fossil fuel use needs to be phased out. There would be no clear-cut scientific basis for this decision. Those taking it would have to base it on an assessment of how big a reduction can be made in the warming effects of the other gases and of how rapidly the land can be turned from an emissions source to an emissions sink as its plants and soils begin to sequester some of the unwanted CO_2 from the air.

Dealing with other causes of warming

Not only does CO_2 have to be removed from the atmosphere but efforts should be made to remove other gases with long atmospheric life-spans – nitrous oxide and the halocarbons – as rapidly as possible so that the rate at which the world is warming actually slows.

"Cutting HFCs, black carbon, tropospheric ozone and methane can buy us about 40 years before we approach the dangerous threshold of 2°C warming." So said Veerabhadran Ramanathan, a distinguished professor of climate and atmospheric sciences at the Scripps Institution of Oceanography, San Diego, in October 2009 at the launch of a paper[38] he co-wrote with Nobel Laureate Mario Molina and others. "If we reduce black carbon emissions worldwide by 50 percent by fully deploying all available emissions-control technologies, we could delay the warming effects of CO_2 by one to two decades and at the same time greatly improve the health of those living in heavily-polluted regions" he added. The paper identifies four "fast-action regulatory strategies" which the authors think could take effect "within 2–3 years and be substantially implemented within 5–10 years, with the goal of producing desired climate response within decades." Besides reducing black carbon, they propose running down the production of hydrofluorocarbons with a high global warming potential, accelerating the phase-out of hydrochlorofluorocarbons and recovering and destroying stratospheric ozone-depleting substances from scrapped products such as refrigerators. However any cooling directly caused by scrapping these powerful greenhouse gases may be at least partially compensated by warming produced by consequent increases in atmospheric ozone, itself a powerful greenhouse gas[39].

Their third proposal is to reduce emissions of pollutants such as carbon monoxide, the nitrogen oxides, methane and volatile organic compounds. These undergo complex photochemical reactions and form ozone in the troposphere which extends up to 15 km above the ground. The paper says that this tropospheric ozone has increased by 30% since pre-industrial times and its contribution to global warming since preindustrial times is as much as 20% of that due to CO_2, in agreement with the values shown in Table 1. In addition, although tropospheric ozone helps to prevent certain forms of skin cancer by absorbing UV radiation, the effect of ground level ozone on human health is strongly negative and in addition, as a 2008 Royal Society report[40] showed, could have caused $26 billion worth of damage to crops in 2000. Reducing ground level ozone by half could delay rising temperatures by another decade, Professor Ramanathan says.

The fourth proposal is to increase the sequestration of CO_2 already in the air through improved forest protection and biochar production. Biochar is charcoal produced from biomass which is ploughed into the soil to retain nutrients, cut the release of methane and nitrous oxide, and increase fertility. Moreover, by encouraging the development of fungi and micro-organisms, it increases the soil's carbon content by much more than its own weight.

The paper cites a study by Lenton and Vaughan[41] which suggests that, under highly optimistic scenarios, the capture of atmospheric CO_2 by plants grown to provide bio-energy followed by capturing and storing the carbon they release when burned combined with afforestation and biochar production may have the potential to remove 100 ppm of CO_2 from the atmosphere. This, alone, would return the atmospheric concentration of CO_2 to near preindustrial levels and reduce the heating effect by 1.3 watts per square metre. "However," the paper says in major understatement, "this may conflict with food production and ecosystem protection".

A UNEP/World Meteorological Organisation report *Integrated Assessment of Black Carbon and Tropospheric Ozone* published in May 2011 reaches much the same conclusions:

Reducing black carbon and tropospheric ozone now will slow the rate of climate change within the first half of this century. Climate benefits from reduced ozone are achieved by reducing emissions of some of its precursors, especially methane which is also a powerful greenhouse gas. A small number of emission reduction measures targeting black carbon and ozone precursors could immediately begin to protect climate, public health, water and food security, and ecosystems. Measures include the recovery of methane from coal, oil and gas extraction and transport, methane capture in waste management, use of clean-burning stoves for residential cooking, diesel particulate filters for vehicles and the banning of field burning of agricultural waste. Widespread implementation is achievable with existing technology

Full implementation of the identified measures would reduce future global warming by 0.5°C (within a range of 0.2–0.7°C.) If the measures were to be implemented by 2030, they could halve the potential increase in global temperature projected for 2050. The rate of regional temperature increase would also be reduced.

Conclusions

This analysis leads me to think that while the climate crisis is alarming it is by no means hopeless provided we break down the problem and the solutions into their component parts and tackle the easiest, rather than the toughest, first. That can buy valuable time. So my suggestions for fellow climate campaigners are as follows.

- Stop being negative. Fossil fuel use may start declining quite soon anyway, climate deal or no climate deal. Talk about the advantages of investing in non-carbon energy sources sooner rather than later, because the energy required for the switch will never be as cheap again in terms of what has to be given up to get it.
- Insist on the highest environmental standards and compensation arrangements being put into place before any further shale gas or underground coal gasification licences are issued. This might be a more productive approach than fighting for an outright ban.
- Campaign for a set of international arrangements that recognises that fossil fuel CO_2 is not the only problem and that a range of programmes is needed to tackle all the causes of warming, even the minor ones.
- As the safe level of greenhouse gases in the atmosphere has already been passed, plants seem at present the only realistic way to extract the excess carbon from the air and sequester it in the soil. So work to prevent further forest loss and to protect the carbon in soils, mires and peat bogs. Advocate re-foresting vast areas, changing grazing methods and using biochar to speed the rate at which long-lived forms of soil carbon can build up. Giving indigenous people an incentive to

maintain woodland and plant more trees for example by encouraging the development of high value sustainable woodland products such a wild silk and honey could help. Millions of people will need to be involved in this effort but increased rural prosperity should result. Moreover the spin-off from new forests could be more abundant water supplies and slower warming as a result of increased cloud cover.

- Talk about the health gains that would come from reducing black carbon emissions and dealing with low-level ozone. Of the rural prosperity that should come from increasing the soil's carbon content and thus its fertility through better farming methods. Of the new, local industries than will spring up making plastics and other organic chemicals out of biomass rather than oil.
- Work to convince people that the global economy will never run properly again unless the benefits of using all scarce resources are fairly shared. Support the setting up of a Global Climate Trust so that the scarcity rent from fossil fuel use gets properly allocated.

Leaving the effects of climate change aside, the main danger that humanity faces is that it will not invest enough of the fossil energy it can extract with a reasonable net-energy gain into making the transition to renewable energy sources. Every post-credit-crunch year that the global economy stays stalled with its engine idling and burning fuel leaves less in the tank to get it anywhere once the clutch is depressed, the gears engaged and a definite direction taken. Future generations will be hungrier, poorer and probably much smaller because of the delay.

Now, for the first time, if we can get people to internalise the implications of oil peak and the other resource constraints, we have a chance to go beyond Lord Stern's mildly-negative position that the cost of dealing with climate change is not very high and move on to the positive position that no costs, and no self-denial are involved. Instead, rationing energy use in order to share out its benefits is essential for the proper working of the economic system and will create millions of jobs and commercial opportunities now.

In particular, campaigners should refute the uber-negative position adopted by Clive Hamilton in his 2010 book *Requiem for a Species* that it is now too late to do anything about the climate except to resign ourselves and die with dignity. The main reason Hamilton thinks this is the case is that he believes that the institutions and thought processes which would need to change to make a better outcome possible will not do so in time to save the day. If, as he assumes, economic growth was still possible, he might be right. The promise of higher incomes for the next few years would almost certainly continue to place an effective block on proposals to cut fossil fuel use, and thus incomes, now. But that block is cleared away by the recognition that regulating fuel demand means much higher incomes in the medium to long term than those that would result if a market-free-for-all led to an economic collapse. Environment and business can walk hand-in-hand and once the limits to the fossil energy supply and thus to economic growth are recognised their interests become aligned.

So there are strong grounds for believing that the climate crisis can be overcome and that many people's lives, particularly in the poorer countries, could be materially better than they are now because of the work the production of biofuels and biochemicals to replace their fossil equivalents should bring, coupled with the additional fertility that biochar should create. Since the alternative is industrial and societal decline and, after increasing unrest, an eventual collapse, there's every reason to think the system will incline the right way. But one thing is necessary first: the twin myths that there's plenty of energy and that economic growth can continue must be exposed. If climate campaigners can get that message over, their battle would be as good as won.

Endnotes

1. IPCC 2007 Climate Change 2007: The Physical Science Basis. Contribution of Working Group 1 to the Fourth Assessment Report of the Intergovernmental Panel on Climate Change ed S Solomon, D Qin, M Manning, Z Chen, M Marquis, K B Avery, M Tignor and H L Miller (Cambridge & New York: Cambridge University Press) 996 pp
2. World Meteorological Organisation press release, 20 January, 2011. http://www.wmo.int/pages/mediacentre/press_releases/pr_906_en.html
3. http://www.tradingeconomics.com/sweden/fossil-fuel-energy-consumption-percent-of-total-wb-data.html
4. In his June 20011 paper The End of Cheap Uranium, Michael Dittmar, of the Institute of Particle Physics, Zurich, Switzerland, says that global uranium mining will peak around 2015 and will then decline with the result that there will not be sufficient to fuel the existing and planned nuclear power plants during the next 10-20 years. Unless nuclear is phased out, he says, "some countries will simply be unable to afford sufficient uranium fuel at that point, which implies involuntary and perhaps chaotic nuclear phase-outs in those countries involving brownouts, blackouts, and worse." See http://xxx.lanl.gov/PS_cache/arxiv/pdf/1106/1106.3617v1.pdf
5. http://www.paulchefurka.ca/WEAP/WEAP.html
6. Zero Carbon Britain (2007) *ZeroCarbonBritain: an alternative energy strategy*, Machynlleth: Centre for Alternative Technology.
7. http://aspoireland.files.wordpress.com/2009/12/newsletter100_200904.pdf
8. "A global coal production forecast with multi-Hubbert cycle analysis" Energy 35 (2010) 3109e3122
9. "Global coal production outlooks based on a logistic model" June 2010 Fuel 89 (2010) 3546–3558
10. The Offshore Valuation Report, PIRC, 2010, pages 60 & 61 http://www.offshorevaluation.org/downloads/offshore_vaulation_full.pdf
11. "The Good News About Coal" June 28, 2011, http://www.earth-policy.org/book_bytes/2011/wotech13_ss3
12. "Preliminary Assessment of the Greenhouse Gas Emissions from Natural Gas obtained by Hydraulic Fracturing" March 2010, http://www.eeb.cornell.edu/howarth/GHG%20emissions%20from%20Marcellus%20Shale%20--%20with%20figure%20--%203.17.2010%20

draft.doc.pdf

13. http://www.usclimatenetwork.org/resource-database/report-coal-to-gas-the-influence-of-methane-leakage
14. Section 2.9.3 Global Mean Radiative Forcing by Emission Precursor, IPCC Fourth Assessment Report (AR4)
15. "Preliminary Assessment of the Greenhouse Gas Emissions from Natural Gas obtained by Hydraulic Fracturing" March 2010, http://www.eeb.cornell.edu/howarth/GHG%20emissions%20from%20Marcellus%20Shale%20--%20with%20figure%20--%203.17.2010%20draft.doc.pdf
16. Broderick. J., et al: 2011, Shale gas: an updated assessment of environmental and climate change impacts. A report commissioned by The Co-operative and undertaken by researchers at the Tyndall Centre, University of Manchester downloaded free from http://www.tyndall.ac.uk/sites/default/files/broderick2011_shalegasexecsummary_conclusions.pdf
17. Private conversation, 9 September, 2010, with the Irish energy minister, Eamon Ryan, who attended the meeting.
18. Sorrell, S. et al (2009) "Global Oil Depletion. An assessment of the evidence for a near-term peak in global oil production", UK Energy Research Centre, Aug. 2009. Copy (3.8 MB pdf): http://www.ukerc.ac.uk/support/tiki-download_file.php?fileId=283
    Aleklett, K., M. Höök, K. Jakobsson, M. Lardelli, S. Snowden & B. Söderbergh, (2010) "The peak of the oil age" Energy Policy, Vol.38 no.3, pp.1398–1414, Mar. 2010 http://www.tsl.uu.se/uhdsg/publications/peakoilage.pdf;
19. Sherwood, S.C. and Huber, M. (2010). "An adaptability limit to climate change due to heat stress". PNAS 107 ,(21), 9552–9555. http:/ www.pnas.org/cgi/doi/10.1073/pnas.0913352107. See also Schmidt, G. and Archer D.. 2009. "Too much of a bad thing" Nature , 458: 1117-1118 downloaded free from http://pubs.giss.nasa.gov/abs/sc00400a.html
20. Page, S.E., F. Siegert, J. O. Rieley, V. Boehm Hans-Dieter, A. Jaya, and S. Limin. (2002). "The amount of carbon released from peat and forest fires in Indonesia during 1997". Nature 420: 61. 65
21. It will not be possible to use the fossil CO2 emissions permits to cover these emissions as, if there was a net loss of biomass and soil carbon in any year, the use of fossil permits to cover its release would mean that there was an inadequate number of permits left for the purchase of that year's capped amount of fossil fuel. If the fossil fuel producers had agreed to participate in the capping scheme on the basis that their annual sales and the income from them were guaranteed, as seems likely, the failure to purchase all their fuel would be in breach of the Trust's agreement with them.
22. http://www.forestcouncil.org/tims_picks/view.php?id=2103
23. Ekstrom, S. et al., (2010) "A possible role of ground-based microorganisms on cloud formation in the atmosphere", Biogeosciences, 7, 387–394, 2010 http://www.biogeosciences.net/7/387/2010/bg-7-387-2010.pdf.
24. "CLAW hypothesis". http://en.wikipedia.org/wiki/CLAW_hypothesis
25. Ravishankara, A.R., M.J. Kurylo, and A.-M. Schmoltner, (2008):."Introduction", In: Trends in Emissions of Ozone-Depleting Substances, Ozone Layer Recovery, and Implications for Ultraviolet Radiation Exposure, NOAA; downloaded from http://www.climatescience.gov/Library/sap/sap2-4/final-report/#finalreport. See also V.E. Fioletov (2008): Ozone climatology, trends, and substances that control ozone, Atmosphere-Ocean,46:1, 39-67 which may be purchased from http://www.tandfonline.com/doi/abs/10.3137/ao.460103?journalCode=tato20.
26. Anon (2011) Report to Congress on Black Carbon External Peer Review Draft http://yosemite.epa.gov/sab/sabproduct.nsf/0/05011472499C2FB28525774A0074DADE/$File/BC%20RTC%20External%20Peer%20Review%20Draft-opt.pdf
27. Skeie, R. B.,. Berntsen, T. K., Myhre, G., Tanaka, K., Kvalev M. M. and Hoyle, C.R. (2011)

"Anthropogenic radiative forcing time series from pre-industrial times until 2010", Atmos. Chem. Phys. Discuss., 11, 22545–22617. http://www.atmos-chem-phys-discuss.net/11/22545/2011/acpd-11-22545-2011.html

28. Woods Hole Research Center, http://www.whrc.org/carbon/index.htm
29. Ravishankara, A.R., M.J. Kurylo, and A.-M. Schmoltner, (2008):."Introduction", In: Trends in Emissions of Ozone-Depleting Substances, Ozone Layer Recovery, and Implications for Ultraviolet Radiation Exposure, NOAA; downloaded from http://downloads.climatescience.gov/sap/sap2-4/sap2-4-final-all.pdf. See also V.E. Fioletov (2008): Ozone climatology, trends, and substances that control ozone, Atmosphere-Ocean,46:1, 39-67 which may be purchased from http://www.tandfonline.com/doi/abs/10.3137/ao.460103?journalCode=tato20
30. http://tamino.wordpress.com/2011/05/28/methane-update/
31. http://tamino.wordpress.com/2011/05/28/methane-update/
32. Anon (2011) Report to Congress on Black Carbon External Peer Review Draft http://yosemite.epa.gov/sab/sabproduct.nsf/0/05011472499C2FB28525774A0074DADE/$File/BC%20RTC%20External%20Peer%20Review%20Draft-opt.pdf
33. "Atmospheric carbon dioxide removal: long-term consequences and commitment", June 2010, http://iopscience.iop.org/1748-9326/5/2/024011/
34. http://theenergycollective.com/TheEnergyCollective/47628
35. Hansen J. et al. (2008) Target atmospheric CO2: Where should humanity aim?, Open Atmos. Sci. J. (2008), vol. 2, pp. 217-231 http://arxiv.org/abs/0804.1126
36. http://www.guardian.co.uk/environment/2008/sep/15/climatechange.carbonemissions
37. "Atmospheric carbon dioxide removal: long-term consequences and commitment", June 2010, http://iopscience.iop.org/1748-9326/5/2/024011/
38. "Reducing abrupt climate change risk using the Montreal Protocol and other regulatory actions to complement cuts in CO2 emissions" by Mario Molinaa, Durwood Zaelkeb,1, K. Madhava Sarmac, Stephen O. Andersend, Veerabhadran Ramanathane,and Donald Kaniaruf, http://www.pnas.org/content/early/2009/10/09/0902568106.full.pdf+html
39. World Meteorological Association (2011). "The state of greenhouse gases in the atmosphere based on global observations through 2011". WMO greenhouse gas bulletin 7, 21st November 2011. http://www.wmo.int/pages/prog/arep/gaw/ghg/documents/GHG_bull_6en.pdf
40. Royal Society (2008) "Ground-level ozone in the 21st century: Future trends, impacts and policy implications". Available at http://royalsociety.org/uploadedFiles/Royal_Society_Content/policy/publications/2008/7925.pdf
41. Lenton TM, Vaughan NE (2009) "The radiative forcing potential of different climate geo-engineering options". Atmos Chem Phys Disc 9:2559–2608

INDEX

Compiled by Nicky Craven

A

B

C

D

E

F

G

H

I

J

K

L

M

N

O

P

R

S

T

U

V

W

Z